Minerals in Thin Section

Minerals
in Thin Section

Second Edition

Dexter Perkins

Department of Geology and Geological Engineering
The University of North Dakota, Grand Forks

Kevin R. Henke

Center for Applied Energy Research
The University of Kentucky, Lexington

PEARSON
Prentice
Hall

Pearson Education, Inc.
Upper Saddle River, New Jersey 07458

Library of Congress Cataloging-in-Publication Data

Perkins, Dexter.
    Minerals in thin section / Dexter Perkins, Kevin R. Henke.— 2nd ed.
        p.   cm.
        ISBN 0-13-142015-1
    1. Thin sections (Geology)   2. Minerals.   I. Henke, Kevin R.   II.
Title.
    QE434 .P47 2004
    549—dc21

                                            2003006659

Executive Editor: *Patrick Lynch*
Assistant Editor: *Melanie Van Benthuysen*
Editorial Assistant: *Sean Hale*
Vice President and Director of Production and Manufacturing, ESM: *David W. Riccardi*
Production Editor: *Patty Donovan*
Director of Creative Services: *Paul Belfanti*
Creative Director: *Carol Anson*
Art Director: *Jayne Conte*
Cover Designer: *Bruce Kenselaar*
Manufacturing Manager: *Trudy Pisciotti*
Manufacturing Buyer: *Lynda Castillo*
Senior Marketing Manager: *Christine Henry*

All photos are courtesy of the authors unless otherwise indicated.

 © 2004, 2000 by Pearson Education, Inc.
Pearson Education, Inc.
Upper Saddle River, New Jersey 07458

The author and publisher of this book have used their best efforts in preparing this book. These efforts include
the development, research, and testing of the theories and programs to determine their effectiveness. The author and publisher make no warranty of any kind, expressed or implied, with regard to these programs or the-
documentation contained in this book. The author and publisher shall not be liable in any event for incidental
or consequential damages in connection with, or arising out of, the furnishing, performance, or use of these
programs.

Printed in the United States of America

ISBN 0-13-142015-1

Pearson Education, Ltd., *London*
Pearson Education Australia Pty. Ltd., *Sydney*
Pearson Education Singapore, Pte. Ltd.
Pearson Education North Asia Ltd., *Hong Kong*
Pearson Education Canada, Inc., *Toronto*
Pearson Educación de Mexico, S.A. de C.V.
Pearson Education—Japan, *Tokyo*
Pearson Education Malaysia, Pte. Ltd.
Pearson Education, *Upper Saddle River, New Jersey*

Contents

Preface

We wrote the first edition of Minerals in Thin Section because we found a need for a concise and straightforward reference book that our students could use in the laboratory. We were pleased to find that other instructors had a similar need.

Many people have made suggestions on ways to make our book more useful. Please keep those suggestions coming! In this, the second edition of *Minerals in Thin Section,* we added a few more details to mineral description, corrected some minor factual and technical errors, and expanded or clarified discussions in a few places where things were not clear. We added additional photos, most of sedimentary rocks in thin section. Most significant, however, is that we added more than 50 line drawings in Part II. The drawings explain, better than words, the relationships between optical properties and crystal shapes. For a few minerals we have included graphs showing the relationship between optical properties and mineral composition.

Many people contributed to this book. In particular we are indebted to Mickey E. Gunter (University of Idaho), Jennifer A. Thomson (Eastern Washington University), and Edward F. Stoddard (North Carolina State University) for help as we prepared the first edition. Subsequent invaluable input came from from James A. Woodhead (Occidental College), Michael J. Walawender (San Diego State University), James A. Grant (University of Minnesota-Duluth), Roderic Brame (Wright State University), Andrew Wulff (University of Iowa), Laura R. Wetzel (Eckerd College), B. Ronald Frost (University of Wyoming), and Kevin L. Shelton (University of Missouri). George B. Perkins helped draft many of the figures.

If you have any comments about this book, please feel free to email the first author at dexter_perkins@und.edu.

About the Authors

Dr. Dexter Perkins received his Ph.D. from the University of Michigan in 1979. He has published over 80 papers and three books. He has had research appointments at the University of Chicago and the Université Blaise Pascal and has been a regular faculty member in the Department of Geology and Geological Engineering at the University of North Dakota for more than 20 years. His current research is focused on mineral equilibria and science education reform.

Kevin R. Henke received his Ph.D. in geology from the University of North Dakota in 1997. He has had research and postdoctoral appointments at Oak Ridge National Laboratory, in the Chemistry Department at North Dakota State University, and in the Chemistry Department at the University of Kentucky. He has also taught in the Department of Geological Sciences at the University of Kentucky. Currently, he is researching the chemistry and environmental impacts of mercury and other heavy metals as an employee of the Center for Applied Energy Research at the University of Kentucky, Lexington.

PART
I
Theoretical Considerations

This book discusses the interaction of minerals and light, and the properties of minerals in thin section. It discusses the most practical aspects of *optical mineralogy,* which is the branch of mineralogy that deals with the optical properties of minerals. A fundamental principle of optical mineralogy is that most minerals, even dark-colored minerals and others that appear opaque in hand specimens, transmit light if we slice them thinly enough. We use a *polarizing microscope* to examine them by *transmitted light microscopy* (Figure 1). We look at small mineral grains (powdered samples) or specially prepared thin sections (0.03-mm-thick specimens of minerals or rocks mounted on glass slides) to determine properties that are otherwise not discernible. Minerals with metallic luster, and a few others, are termed *opaque minerals.* They don't transmit light even if they are thin-section thickness. For these minerals, transmitted light microscopy is of no use. *Reflected light microscopy,* a related technique, can reveal some of the same properties. It is an important technique for economic geologists who deal with metallic ores but is not used by most mineralogists or petrologists, so we discuss it only briefly in this book.

Most minerals can be identified when examined with a polarizing microscope, even if unidentifiable in hand specimen. Optical properties also allow a mineralogist to estimate the composition of some minerals. For example, the Fe:Mg ratio of olivine [$(Fe,Mg)_2SiO_4$] can be distinguished based on optical properties. Plagioclase feldspar [$CaAl_2Si_2O_8$–$NaAlSi_3O_8$] composition can be similarly distinguished. Box 1 (on the inside front cover of this book) summarizes the optical properties used for mineral identification and gives the properties of some common minerals. We divide minerals into those that will not transmit light unless the sections are much thinner than normal thin sections (*opaque minerals*) and those that will (*nonopaque minerals*). Nonopaque minerals are further divided into those that are *isotropic* (having the same properties in all directions) and those that are *anisotropic* (having different optical properties in different directions). Finally, the anisotropic minerals are divided according to whether they are *uniaxial* or *biaxial,* and according to whether they have a positive or negative *optic sign.* The

details of these, and other diagnostic properties, will be discussed later.

Besides mineral identification, the polarizing microscope reveals important information about rock-forming processes (*petrogenesis*). When we use thin sections, distinguishing igneous, sedimentary, and metamorphic rocks is often easier than when we use hand specimens. More significantly, it is possible to identify minerals and distinguish among different types of igneous, sedimentary, and metamorphic rocks. The microscope allows us to see textural relationships in a specimen that give clues about when and how different minerals in the rock formed. Microscopic relationships between mineral grains allow us to determine the order in which minerals crystallized from magma, and we can identify minerals produced by alteration or weathering long after magma cooling. Similar observations are possible for sedimentary or metamorphic rocks. Only the microscope can give us such information, information that is essential if rocks are to be used to interpret geological processes and environments.

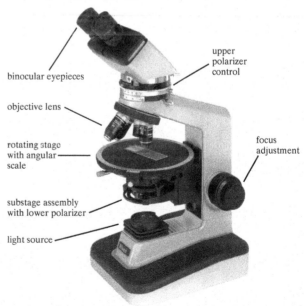

▶FIGURE 1

A polarizing microscope with main features labeled. From Nikon, Inc., Melville, New York. Photo used with permission.

What Is Light?

◤ THE PROPERTIES OF LIGHT

Before starting a discussion of optical mineralogy, it is helpful to take a close look at light and its properties. Light is one form of *electromagnetic radiation* (Figure 2). Radio waves, ultraviolet light, and X rays are other forms of electromagnetic radiation. All consist of propagating (moving through space) electric and magnetic waves. The interactions between electric waves and crystals are normally much stronger than the interactions between magnetic waves and crystals (unless the crystals are metallic). Consequently, this book only discusses the electrical waves of light. In principle, however, much of the discussion applies to the magnetic waves as well. Light waves, like all electromagnetic radiation, are characterized by a particular wavelength, λ, a frequency, v, and a polarization state (Figure 3). The velocity, v, of the wave is the product of λ and v:

$$\mathrm{v} = \lambda v \qquad (1)$$

λ ↑ ν

high energy

λ (wavelength, meters)		ν (frequency, cycles per second)
10^{-16}	gamma rays	10^{24}
10^{-14}		10^{22}
10^{-12}	cosmic rays	10^{20}
1 Å — 10^{-10}	*diameter of atom*	10^{18}
1 nm	X rays / ultraviolet light	
10^{-8}		10^{16}
1 μ — 10^{-6}	visible light	10^{14}
10^{-4}	infrared radiation	10^{12}
1 mm 10^{-2}		10^{10}
	short radio waves	
1 m — 10^{0}		10^{8}
10^{2}	*diameter of a baseball*	10^{6}
1 km 10^{4}		10^{4}
10^{6}	long radio waves	10^{2}
10^{8}	*diameter of the Earth*	10^{0}

low energy

wavelength (meters)   frequency (cycles per second)

►FIGURE 2
The electromagnetic spectrum. Visible light is a form of electromagnetic radiation with wavelengths and energies that fall in the middle of the spectrum.

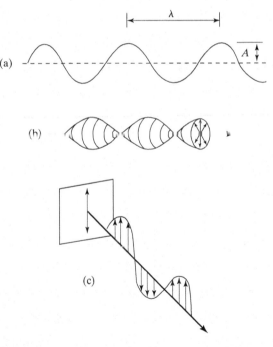

►FIGURE 3
Properties of light rays. (a) Different colors of light are characterized by different wavelengths (λ). The intensity of a wave is proportional to its amplitude (A). (b) The electric vectors of unpolarized light (arrows) vibrate in all directions perpendicular to the direction of travel. (c) The electric vectors of plane polarized light are constrained to vibrate in a plane.

In a vacuum, the velocity of light, c, is 3×10^8 meters per second. Light velocity is slightly less when passing through air, and can be considerably less when passing through crystals. When the velocity of light is altered as it passes from one medium (for example, air) to another (perhaps a mineral), the wavelength changes, but the frequency remains the same.

Visible light has wavelengths of 390 nm to 770 nanometers, which is equivalent to 3,900 to 7,700 Ångstroms, or $10^{-6.1}$ to $10^{-6.4}$ meters. Different wavelengths correspond to different colors of light (Figure 4). The shortest wavelengths, corresponding to violet light, grade into invisible ultraviolet radiation. The longest wavelengths, corresponding to red light, grade into invisible infrared radiation. Light composed of multiple wavelengths appears as one color to the human eye. If wavelengths corresponding to all the primary colors are present with nearly equal intensities, the light appears white. White light is *polychromatic* (many colored), containing a range, or spectrum, of wavelengths. Polychromatic light can be separated into different wavelengths in many ways. When one wavelength is isolated, the light is *monochromatic* (single colored).

▶INTERFERENCE

Besides λ and ν, an amplitude and a phase characterize all waves. *Amplitude* (*A*) refers to the height of a wave. *Phase* refers to whether a wave is moving up or down at a particular time. If two waves move up and down at the same time, they are *in phase;* if not, they are *out of phase.* When two waves interact, traveling in the same direction simultaneously, they interfere with each other. The nature of the interference depends on the relationships between their wavelengths,

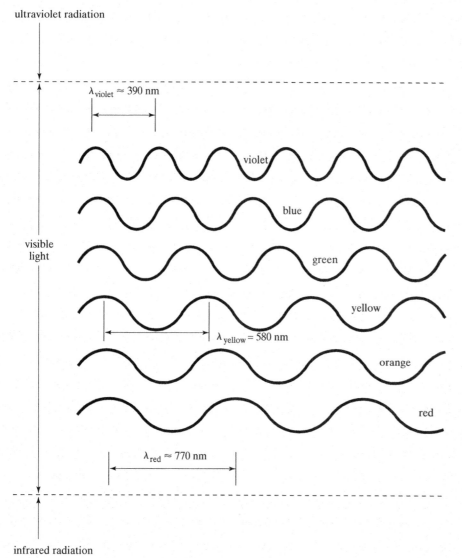

ultraviolet radiation

$\lambda_{\text{violet}} \approx 390$ nm

violet

blue

green

yellow

$\lambda_{\text{yellow}} = 580$ nm

orange

red

$\lambda_{\text{red}} \approx 770$ nm

visible light

infrared radiation

▶**FIGURE 4**
The wavelengths of visible light. The wavelength of violet light is about half that of red light. The boundaries between visible light and invisible radiation are not precisely defined, but visible light grades into ultraviolet radiation at short wavelengths and into infrared radiation at long wavelengths.

amplitudes, and phases. Light waves passing through crystals can have a variety of wavelengths, amplitudes, and phases that are affected by atomic structure in different ways. They yield interference phenomena, giving minerals distinctive optical properties.

In Figure 5a, two in-phase waves of the same wavelength are going in the same direction. If we could measure the intensity of the two waves together, we would find that it is about twice the intensity of each individual wave. When waves are in phase, no energy is lost; this is *constructive interference.* In contrast, Figure 5b shows two waves that are partially out of phase, and Figure 5c shows two waves that are completely out of phase. When waves are out of phase, wave peaks and valleys do not correspond. If they are completely out of phase, the peaks of one wave correspond to the valleys of the other. Consequently, addition of out-of-phase waves can result in *destructive interference,* a condition in which the waves "consume" some or all of each other's energy. For perfect constructive or destructive interference to occur, waves must be of the same wavelength. Interaction of waves with different wavelengths is more complicated.

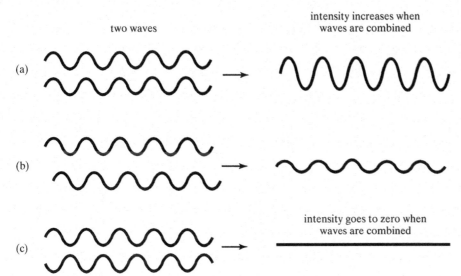

▶FIGURE 5
Waves in phase and out of phase, and the results when they combine. (a) Waves are in phase if their peaks and wavelengths correspond, so they add to produce one wave with twice the amplitude. (b) When waves are partially in phase, their peaks do not exactly correspond; so combination results in some energy loss. (c) When waves are completely out of phase, their motions cancel and addition leads to complete loss of energy.

Polarization of Light and the Polarizing Microscope

▶POLARIZED LIGHT

The vibration motion of a light wave is perpendicular, or nearly perpendicular, to the direction it is propagating. In normal unpolarized beams of light, waves vibrate in many different directions, shown by arrows in Figure 3b. However, we can filter or alter a light beam to make all the waves vibrate in one direction parallel to a particular plane (shown by arrows in Figure 3c). The light is then *plane-polarized,* sometimes called just *polarized.* Light becomes polarized in different ways. Reflection from a shiny surface can partially or completely polarize light because light vibrating in planes parallel to the reflecting surface is especially well reflected, while light vibrating in other directions is absorbed. This is why sunglasses with polarizing lenses help eliminate glare.

Suppose light passes through a polarizing filter that constrains it to vibrate in a north-south (up-down) direction. The polarized beam, although perhaps decreased in intensity, appears the same to our eyes, because human eyes cannot determine whether light is polarized. If, however, another polarizing filter is in the path of the beam, we can easily determine that the beam is polarized (Figure 6). If the second filter allows only light vibrating in a north-south direction to pass, the polarized beam will pass through it (Figure 6a). If we slowly rotate the second filter to an east-west direction, it will gradually transmit less light, and eventually no light (Figure 6b).

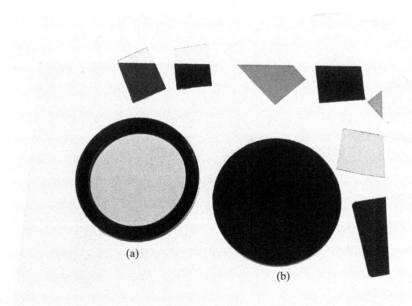

(a)

(b)

▶FIGURE 6
Several small polarizing filters on top of a large polarizing sheet. The amount of light transmitted depends on the relative orientations of the polarization of the sheet and the small filter. (a) When polarization directions of the two are parallel, the maximum amount of light possible is transmitted. (b) When polarization directions are perpendicular, no light is transmitted. At other orientations the two filters transmit intermediate amounts of light.

▶POLARIZING MICROSCOPES

Polarizing microscopes, also called *petrographic microscopes,* are in many respects the same as other microscopes (Figures 1 and 7). They magnify small objects so we can see them in greater detail. A bulb provides a white light source. The light passes through several filters and diaphragms before it reaches the stage and interacts with the material being observed. One of the most important filters is the lower polarizer, which ensures that all light striking samples on the stage is plane polarized (vibrating, or having wave motion, in only one plane). The presence of a lower polarizer sets polarizing microscopes apart from others. In most modern polarizing microscopes, the lower polarizer only allows light vibrating in an east-west direction to reach the stage. Older microscopes, however, have the lower polarizer oriented in a north-south direction. A fixed condensing lens and a diaphragm in the substage help concentrate light on the sample. For most purposes, we use *orthoscopic illumination;* in which an unfocused beam travels from the substage through the sample and straight up the microscope tube. The light rays travel orthogonal to the stage and to a sample or thin section on the stage. However, we can insert a special lens, a *conoscopic lens,* between the lower polarizer and stage to produce *conoscopic illumination* when needed (Figure 7). The conoscopic lens, also called a *condenser lens,* causes the light beam to converge (focus) on a small spot on the sample and illuminates the sample with a cone of non-parallel rays.

We can rotate the microscope stage to change the orientation of the sample relative to the polarized light. Because most minerals are anisotropic, the in-

Orthoscopic Illumination   **Conoscopic Illumination**

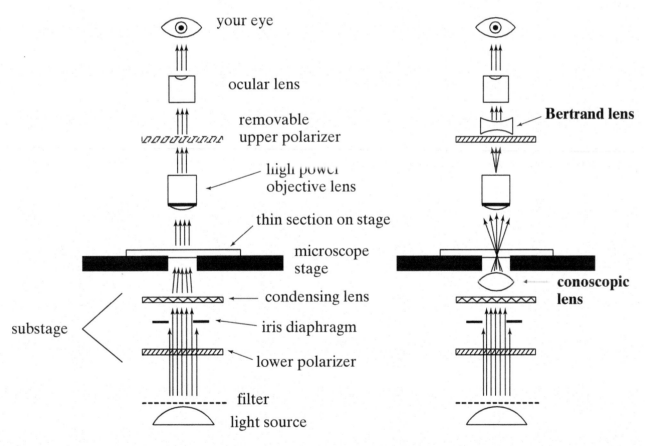

▶FIGURE 7

The most important components of a polarizing microscope. Light from a bulb passes through a filter, the lower polarizer, a diaphragm, and a condensing lens in the substage before it hits the sample on the stage. Above the stage the objective and ocular lenses magnify and focus the light. The upper polarizer, Bertrand lens, and conoscopic lens are inserted when needed.

teraction of the light with a mineral varies with stage rotation. A calibrated angular scale allows us to make precise measurements of crystal orientation. The scale is also useful for measuring angles between cleavages, crystal faces, and twin orientations, and for measuring other optical properties.

Above the stage, a rotating turret holds several *objective lenses.* They usually range in magnification from about 2X to 50X. Different objective lenses can have different *numerical apertures* (N.A.), a value that describes the angles at which light can enter a lens, which is an important consideration when making certain measurements. *In the discussion of interference figures later in this book, we have assumed that the objective lens being used has an N.A. of 0.85, since this is by far the most common today. If you use a lens with a different N.A., some of the given angular values may be in error.* The *ocular,* an additional lens usually providing 8X or 10X magnification, is in the eyepiece. Binocular microscopes, such as the one in Figure 1, have two eyepieces and two oculars. Oculars have cross hairs that aid in making angular measurements when we rotate the stage. The total magnification, the product of the objective lens magnification and the ocular magnification, varies from about 16X to 500X, depending on the lenses used.

We can insert several other filters and lenses between the objective lens and the ocular when needed (Figure 7). The *upper polarizer,* sometimes called the *analyzer,* is a polarizing filter oriented at 90° to the lower polarizer, which we can insert or remove from the path of the light beam. If no sample is on the stage, light that passes through the lower polarizer cannot pass through the upper polarizer. If a sample is on the stage, it usually changes the polarization of the light so that some can pass through the upper polarizer. We can also insert an *accessory plate* above the upper polarizer. The most common kind of accessory plate used today is called a "full wave" plate. In the past, all full wave plates were made of gypsum and are still often referred to as "gypsum plates," but today they are made of quartz. Above the accessory plate, most polarizing microscopes have a *Bertrand lens* and diaphragm. We use them with the substage conoscopic lens to view minerals in conoscopic illumination, allowing us to make some special kinds of measurements.

Petrologists and mineralogists use polarizing microscopes with or without the upper polarizer inserted (Box 2; inside back cover). Without the upper polarizer, we see the sample in *plane polarized light* (*PP* light); with the upper polarizer, we see it in *crossed polars* (*XP* light). Grain size, shape, color, cleavage, and other physical properties are best revealed in PP light. The optical properties refractive index and

pleochroism are also determined using PP light. We use XP light, sometimes focused with conoscopic and Bertrand lenses, to determine properties including retardation, optic sign, and *2V.* These properties are discussed in detail later in this book.

We examine minerals or rocks in *grain mounts* or in *thin sections* (Figure 8). For determining some mineral properties, a small amount of a powdered mineral sample is placed on a glass slide to produce a grain mount. The grains must be thin enough so that light can pass through them without a significant loss of intensity, usually 0.10 to 0.15 mm in longest dimension. A small amount of liquid (often referred to as a *refractive index oil*) surrounds them, and a thin piece of glass, called a cover slip, is placed over the grains and liquid. Grain mounts and refractive index oils are absolutely necessary for making some types of measurements. Petrologists use thin sections, however, for routine mineral identification and other petrographic work. For those interested in more information about studying minerals in grain mounts, see the optical mineralogy texts listed in the references on page 30.

►COLORS IN PLANE POLARIZED (PP) LIGHT AND CROSSED POLARIZED (XP) LIGHT

In hand specimens, many minerals appear strongly colored, but minerals viewed with a microscope using PP light generally display a weak color or appear colorless. Many minerals in a thin section or in a grain

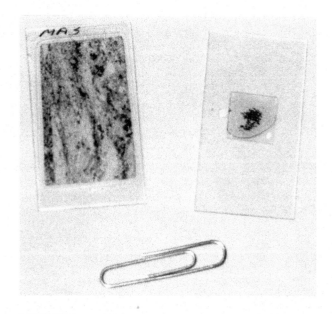

►FIGURE 8
Thin section and grain mount

mount are not thick enough to absorb significantly or enhance specific wavelengths of light. If minerals do appear colored, the color may change when we rotate the microscope stage, because rotating the stage changes the orientation of the mineral's crystal structure with respect to the polarized light. Some minerals absorb different wavelengths of light depending on light vibration direction. We call this property *pleochroism*. Biotite is an example of a mineral that normally displays marked pleochroism. Plate 18c shows biotite grains in different orientations; the color of this biotite varies from greenish-brown to brown. Some biotites in other rocks are even more pleochroic.

Pleochroism is an especially useful diagnostic property when identifying some minerals, but it can be overlooked. In thin sections, orthopyroxenes are commonly colorless, but some show a faint pleochroism from pink to green (Plate 21c). Pleochroism of pyroxenes is an important property because it distinguishes the two major pyroxene subgroups: orthopyroxene and clinopyroxene. For minerals with noticeable pleochroism, reference tables describe the property by listing colors seen when looking at the mineral in different directions. For pleochroic uniaxial minerals, color varies between two hues. For biaxial minerals, color varies between three hues.

In contrast with pyroxenes, many amphiboles display strong colors and a very noticeable pleochroism in a thin section. The biaxial mineral glaucophane (an amphibole) has pleochroism described by its *pleochroic formula:*

X = colorless or pale blue
Y = lavender-blue or bluish green
Z = blue, greenish blue, or violet

X, Y, and Z refer to light vibrating parallel to each of three mutually perpendicular vibration directions in the crystal. In thin sections, glaucophane's colors vary within the limits described for X, Y, and Z, depending on the crystal orientation, as we rotate the microscope stage (Plate 22e). The biotite in Plate 18c is pleochroic in green and brown, but the standard pleochroic formula for biotite might be:

X = colorless, light tan, pale greenish brown, or pale green
Y ≅ Z = brown, olive brown, dark green, or dark red-brown

When we insert the upper polarizer, we see minerals in *crossed polarized* (*XP*) light, and we may see colors that are brighter and more pronounced than when we view the same grain in **PP** light. These are *interference colors.* They do not result from absorption of different wavelengths by the mineral (which is how minerals get their normal color). Instead, they result from the interference of light rays passing through the upper polarizer. They rarely resemble the true color of the mineral. Interference colors depend on grain orientation, so different grains of the same mineral in one thin section normally display a range of interference colors. Because different minerals can display different ranges of interference colors, interference colors are useful for mineral identification. Interference colors also vary with the thickness of the grains, so it is important that thin sections be of uniform thickness. Additionally, the edges of some grains, grains near the edge of a thin section, or grains adjacent to holes in a thin section (places where the sample is thin), may display abnormal interference colors.

The Velocity of Light in Crystals and the Refractive Index

When electromagnetic radiation passes near an atom, the electric wave causes electrons to oscillate. The oscillations absorb energy from the light, and the wave slows down. A wave's velocity through a crystal is described by the crystal's refractive index (n), which depends on chemical composition, crystal structure, and bond type in the crystal. The refractive index (n) is the ratio of the velocity (v) of light in a vacuum to the velocity in the crystal:

$$n = v_{vacuum} / v_{crystal} \qquad (2)$$

Because light passes through a vacuum faster than through any other medium, n always has a value greater than 1. High values of n correspond to materials that transmit light slowly. Under normal conditions, the refractive index of air is 1.00029. Because it is much easier to work with air than with a vacuum, this is a common reference value.

As light passes from air into most nonopaque minerals, its velocity decreases by a third or a half. Because the frequency of the light remains unchanged, we know that the wavelength must decrease by a similar fraction (Equation 1; see page 3). Most minerals have refractive indices between 1.5 and 2.0. Fluorite, borax, and sodalite are examples of minerals that have a very low (< 1.5) index of refraction. At the other extreme, zincite, diamond, and rutile have very high indices (> 2.0). The refractive index is one of the most useful properties for identifying minerals in a grain mount but is less valuable when we examine minerals in thin sections (Plate 1).

The refractive index of most materials varies with the wavelength of light. In other words, the velocity of light in a crystal varies with the light's color. This property, called *dispersion,* is a property of minerals that can sometimes be seen in thin sections but is not dis-

cussed in detail in this book. An excellent but non-mineralogical example of dispersion is the separation of white light into colored "rainbows" when refracted by a glass prism. When a beam of white light enters a prism, different wavelengths (colors) are refracted at different angles, resulting in the production of the "rainbow." For a mineralogical example, we may consider diamond. Diamond's extreme dispersion accounts, in part, for the play of colors ("fire") that diamonds display. Minerals with low dispersion, such as fluorite, appear dull no matter how well cut or faceted. They may, however, be useful as lenses when dispersion causes unwanted effects.

A mineral's refractive index and dispersion profoundly affect its *luster.* Minerals with a very high refractive index and dispersion, such as diamond or cuprite, appear to sparkle and are termed *adamantine.* Minerals with a moderate refractive index, such as spinel and garnet ($n = 1.5 - 1.8$), may appear vitreous (glassy) or shiny, while those with a low refractive index, such as borax, will appear drab because they do not reflect or refract as much incident light. Refractive index depends on many things, but a high n value suggests minerals composed of atoms with high atomic numbers, or of atoms packed closely together.

Most minerals are *anisotropic,* so their refractive index varies with direction. In contrast, a glass, such as window glass or obsidian, is *isotropic* because it has a random atomic structure. Randomness means that, on the average, the structure and refractive index are the same in all directions. Isotropic minerals are relatively easy to spot in thin sections. When viewed with a polarizing microscope and XP light, they remain *extinct,* appearing black as the stage rotates, no matter what their orientation is on the microscope stage. The most common isotropic minerals are listed in Appendix B; the list includes such common minerals as garnet, sphalerite, and fluorite. Sometimes thin sections contain holes, places with no mineral and only epoxy. They appear isotropic and can occasionally be mistaken for isotropic minerals (see Plates 1e–f, 20e–f, 30a–b, 33e–f). We can normally tell isotropic minerals apart by looking at color, relief, habit, and cleavage.

Anisotropic minerals normally do not appear extinct under XP light, but, as the microscope stage rotates, they go extinct briefly every 90°. However, if we orient an anisotropic crystal so that light passes through it parallel to a special direction called an *optic axis*, it will appear isotropic. It remains extinct when we rotate the stage. Fortunately, anisotropic minerals can only have one (*uniaxial minerals*) or two (*biaxial minerals*) optic axes, so the odds of the optic axis being exactly parallel to the light beam are small, and confusing isotropic and anisotropic minerals is rarely a problem. When in doubt, we can distinguish them using conoscopic illumination because anisotropic minerals will transmit some conoscopic light and display interference figures (discussed later), while isotropic minerals do not.

▶SNELL'S LAW AND LIGHT REFRACTION

We have all seen objects that appear to bend as they pass from air into water. A straw in a glass of soda, or an oar in the water, seem bent when we know they are not. We call this phenomenon *refraction*. Refraction occurs when a beam of light passes from one medium to another with a different refractive index (Figure 9). If the light strikes the interface at an angle other than 90°, it changes direction. Consider a beam traveling from air into water (Figure 9a, b). The side of the beam that reaches the interface first will be slowed as it enters the water. The beam bends toward the water, the medium with a higher refractive index, because one side of the beam moves faster than the other. Figure 9c and 9d show the opposite case: a beam traveling from a medium with a high refractive index to another with a lower refractive index. The beam refracts toward the medium with a higher index, as in Figure 9a. A beam traveling at 90° to an interface, whether from a medium with a high refractive index to a low or vice versa, is not refracted at all.

The angle between the incoming beam and a perpendicular to the interface is the *angle of incidence* (θ_i). The angle between the outgoing beam and a perpendicular to the interface is the *angle of refraction* (θ_r). The relationship between the angle of incidence (θ_i) and the angle of refraction (θ_r) is

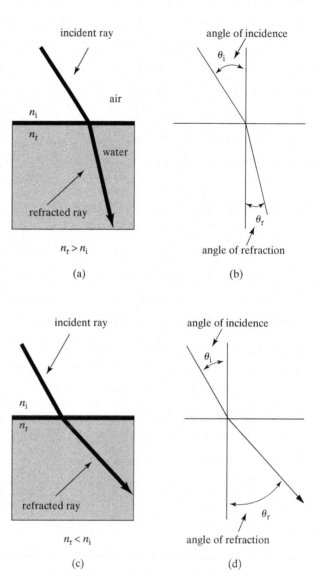

▶FIGURE 9
Refraction of a beam of light. (a) A light ray is bent as it crosses the boundary from air into water (or from any medium to another with higher refractive index). (b) The geometry of refraction shown in (a). (c) A light ray is bent as it crosses the boundary from one medium to another with lower refractive index. (d) The geometry of refraction shown in (c).

$$\sin \theta_i / \sin \theta_r = v_i / v_r = n_r / n_i \qquad (3)$$

where v_i and v_r are the velocities of light through two media, and n_r and n_i are the indices of refraction of the two media. This relationship, *Snell's Law*, is named after Willebrod Snell, the Dutch scientist who first derived it in 1621.

Rearranging Snell's Law tells us that we can calculate the angle of refraction:

$$\theta_r = \sin^{-1} [n_i / n_r \times \sin \theta_i] \qquad (4)$$

By definition, sin values can never be greater than 1.0. Suppose a light beam is traveling from a crystal into air. In this case, $n_i > n_r$ and, because the term in square brackets on the right-hand side of Equation 4 must be less than or equal to 1.0, for some large values of θ_i there is no solution. The limiting value of θ_i is the *critical angle* of refraction. If the angle of incidence is greater, none of the light will escape; the entire beam will be reflected inside the crystal (Figure 10). This is the reason crystals with a high refractive index, such as diamond, exhibit internal reflection that gives them a sparkling appearance. Measuring the critical angle of refraction is a common method for determining refractive index of a mineral. Instruments called *refractometers* simplify such measurements.

▶RELIEF AND BECKE LINES

If we immerse an isotropic mineral grain in a liquid with the same refractive index, we will have difficulty seeing it unless it is one of the few minerals with very strong coloration. The edges of the mineral grain will not stand out. However, if a grain has an index of refraction that is significantly different from the liquid, light refracts and reflects at the edges of the grains. As the difference between the index of the liquid and the mineral increases, the boundary between the two becomes more pronounced (Plate 1). The term *relief* describes the contrast between the mineral and its surroundings (in this case, liquid). Grains with low relief are barely visible, while those with high relief stand out clearly (Figure 11; Plate 1).

Minerals in thin sections also show relief (Plate 1d). The relief depends on the difference in the indices of refraction of the mineral and the material (today usually a special type of Epoxy) in which it is mount-

ed. As the difference in indices increases, relief becomes more noticeable. Minerals with high refractive indices show high ("positive") relief because their index of refraction is *greater* than that of the Epoxy (such as the garnet in Plate 1d). They also tend to show structural flaws, such as scratches, cracks, or pits, more than those with low refractive indices. Some minerals (fluorite, for example) with very low refractive indices also show high relief (termed "negative" relief) because their index of refraction is *lower* than that of the Epoxy. We do not differentiate between positive and negative relief in this book; for most purposes, we need only to know whether a mineral displays high, medium, or low relief (Plate 1a–d). A few minerals (such as calcite) display variable relief with stage rotation; variable relief is a useful diagnostic property. Plate 1d shows garnet (high index of refraction) surrounded by quartz (much lower index of refraction). Garnet has much greater relief and so stands out more. Similarly, the sillimanite in Plate 29e shows very high relief. We can see relief with either a monocular or a binocular microscope, but more easily with the latter.

As pointed out before, when we immerse a grain in liquid, some light rays bend toward the medium with the refractive higher index. Other light rays are completely reflected because they hit the mineral-liquid interface at an angle greater than the critical angle of refraction. The light interacts with the grain as if it were a small lens (Figure 12). If $n_{mineral} > n_{liquid}$, light rays are refracted and converge after passing through the grain. If $n_{mineral} < n_{liquid}$, light rays are refracted and diverge after passing through the grain. If we slowly lower the microscope stage, shifting the focus to a point above the mineral grain, a bright narrow band of light called a *Becke line* appears at the interface and moves toward the material with higher refractive index (Figure 12; Plate 1a–c). A complementary, but more difficult to see, dark band moves

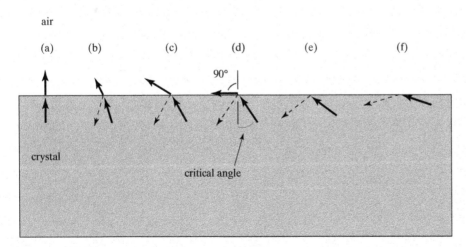

▶**FIGURE 10**
Refraction and reflection of a beam of light passing from a crystal into air for various angles of incidence. The solid rays show the incident and refracted beams; the dashed rays are the reflected beams. In (a), (b), and (c), the angle of incidence is less than the critical angle, so a refracted beam escapes the crystal. In (e) and (f), the angle of incidence is greater than the critical angle, so all the light is reflected back into the crystal.

▶**FIGURE 11**
Minerals appear to have high relief when in a liquid if their index of refraction differs greatly from that of the liquid. Minerals have low relief, and may nearly disappear, if the mineral and the liquid have similar indices of refraction. This figure shows grossular (low relief and hard to see, $n = 1.750$) and fluorite (high relief, $n = 1.430$) in a liquid with $n = 1.720$.

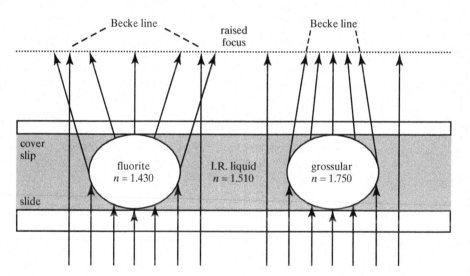

▶**FIGURE 12**
Mineral grains behave like lenses when immersed in a liquid of a different index of refraction, causing light to converge or diverge. In this photo, the fluorite grains (left) cause light to diverge and the grossular grains (right) cause light to converge because $n_{fluorite} < n_{liquid} < n_{grossular}$. Consequently, if the microscope stage is lowered so the focus is raised to a plane above the grains, narrow bright Becke lines will move out into the liquid from the fluorite-liquid boundary, and in toward the center of the grain from the grossular-liquid boundary. When the stage is lowered, Becke lines always move toward the crystal or liquid with the greater index of refraction.

toward the material with lower refractive index. Although not as straightforward, we can also use Becke lines to compare the relief of minerals in thin sections by purposely focusing and defocusing the microscope while we examine a grain boundary. We also compare relief by noting how well a mineral appears to stand out above another (Plate 1d).

Interaction of Light and Crystals

▶ DOUBLE REFRACTION

In most modern polarizing microscopes, polarized light leaves the lower polarizer vibrating in the east-west direction. If it encounters an isotropic mineral on the stage, it slows as it passes through the mineral, but is still east-west polarized when it emerges. Upon entering an anisotropic crystal, however, light is normally split into two polarized rays, each traveling through the crystal along a slightly different path with a slightly different velocity and refractive index (Figure 13a). For uniaxial minerals, we call the two rays the *ordinary ray* (O ray), symbolized by ω, and the *extraordinary ray* (E ray), symbolized by ε. The O ray travels a path predicted by Snell's Law, while the E ray does not. The O ray and E ray vibration directions depends on the direction the light is traveling through the crystal structure, but the vibration directions of the two rays are always perpendicular to each other (Figure 13b, c).

We call the splitting of a light beam into two perpendicularly polarized rays *double refraction*. All randomly oriented anisotropic minerals cause double refraction. We can easily observe it by placing clear calcite over a piece of paper on which a line, dot, or other image has been drawn (Figure 14). Two images appear, one corresponding to each of the two rays. A thin piece of polarizing film placed over the calcite crystal would verify that the two rays are polarized and vibrating perpendicular to each other. If we rotate either the film or the crystal, every 90° one ray becomes extinct, and we will see only one image. Calcite is one of the few common minerals that exhibits double refraction that is easily seen without a microscope, but even minerals that exhibit more subtle double refraction can be tested using polarizing filters. Gemologists use this technique to tell

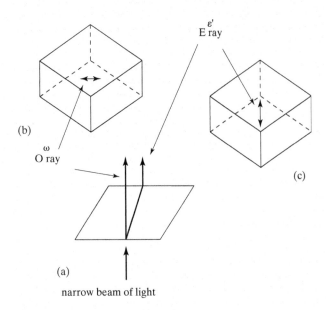

▶ **FIGURE 13**
Splitting of polarized light into an ordinary (O) and extraordinary (E) ray by calcite. (a) A side view showing that the paths of the E ray and the O ray are different. (b) The O ray vibrates parallel to the long axis of the rhomb. (c) The E ray vibrates perpendicular to the long axis of the rhomb.

▶ **FIGURE 14**
Calcite crystal showing double refraction

gems from imitations made of glass. Glass, like all isotropic substances, does not exhibit double refraction.

As the two rays pass through an anisotropic crystal, they travel at different velocities unless they are traveling parallel to an *optic axis.* We call the two rays the *slow ray* and the *fast ray.* Because the rays travel at different velocities, their refractive indices must be different. The difference in the refractive indices of the fast ray and the slow ray ($n_{slow} - n_{fast}$) is the *apparent birefringence* (δ'). It varies depending on the direction light is traveling through the crystal and ranges from zero to some maximum value (δ) determined by the crystal structure. The maximum birefringence (δ) is a diagnostic property of minerals.

When the slow ray emerges from an anisotropic crystal, the fast ray has already emerged and traveled some distance. This distance is the *retardation,* Δ. Retardation is proportional to both the thickness (t) of the crystal and to the birefringence in the direction the light is traveling (δ'):

$$\Delta = t \times \delta' = t \times (n_s - n_f) \qquad (5)$$

The birefringence and retardation of isotropic crystals are always zero. No double refraction occurs, and all light passes through isotropic crystals with the same velocity because the refractive index is equal in all directions. Most anisotropic minerals have birefringence between 0.01 and 0.20. Appendix E contains a list of minerals ordered by interference colors, which are a function of birefringence. Table 3, in Part II of this book, is an abridged version containing only relatively common minerals.

▶ CRYSTALS BETWEEN CROSSED POLARS

When viewed with the upper polarizer in place (under crossed polar (XP) light), we can differentiate isotropic and anisotropic crystals. Suppose we are viewing an isotropic crystal using XP light. It will remain dark through 360° of stage rotation. (See, for example, the garnets in Plates 28d and 28f.) This is because the light emerging from the mineral retains the polarization it had on entering and will always be east-west polarized. It cannot pass through the upper polarizer, oriented at 90° to the lower polarizer. The effect is the same as if no mineral was on the stage.

When we view an anisotropic crystal with XP light, light is split into two rays unless we are looking down an optic axis. The two rays, after emerging from the crystal, travel on to the upper polarizer where they are resolved into one ray with north-south polarization. Because the vibration directions of both the rays are normally not perpendicular to the upper polarizer, components of both pass through the upper polarizer and combine to produce the light reaching our eye. As we rotate the microscope stage, however, the relative intensities of the two rays emerging from the crystal vary. Every 90°, the intensity of one is zero, and the other is vibrating parallel to the lower polarizer. Consequently, no light passes through the upper polarizer and the crystal appears extinct every 90°.

If we used a monochromatic light source in our microscope and looked at an anisotropic crystal under XP light, it would go from light to complete darkness as we rotate the stage. Extinction would occur every 90°, and maximum brightness would be at 45° to the extinction positions. However, most polarizing microscopes use polychromatic light. Because of dispersion, double refraction is slightly different for different wavelengths. Minerals with high dispersion may never appear completely dark, but most come close.

▶ INTERFERENCE COLORS

When white light passes through an anisotropic mineral, all wavelengths are split into two polarized rays vibrating at 90° to each other. Different colors have different wavelengths, so when the rays leave the crystal, some colors may be retarded an even number of wavelengths, but most will not. Consequently, when the north-south components of the two rays are combined at the upper polarizer, constructive interference occurs for some colors, and destructive interference for others. If we look at a mineral of uniform thickness under XP light, we see one color, the *interference color.* Interference colors depend on the retardation of different wavelengths, which in turn depends on the orientation, birefringence, and thickness of a crystal. Interference colors change intensity and hue as we rotate the stage; they disappear every 90°, when the mineral goes extinct.

Normal interference colors are shown in a *Michel-Lévy Color Chart* (see Plate 2). Very low-order interference colors, corresponding to a retardation of less than 200 nm, are gray and white. The interference color of a mineral with very low birefringence, then, changes from white (or gray) to black every 90° as we rotate the microscope stage. For minerals with slightly greater birefringence, yellow, orange, or red interference colors will appear when we rotate the stage. These colors, corresponding to retardation of 200 nm to 550 nm, are called *first order* colors. As retardation increases further, colors repeat every 550 nm. They go from violet to red (*second order*) and then from violet to red again (*third order*). They become more pastel (washed out) in appearance as order increases.

Fourth order colors are often so weak that they appear *"pearl" white* and may occasionally be confused with first order white (see Plate 8). When describing an interference color, it is important to state both the color and the order; for mineral identification, the order is often more important than the color. 550 nm, the difference in retardation between "orders" is the average wavelength of visible light.

Plate 8b shows calcite. The view between crossed polars shows little color, but calcite has extreme birefringence ($\delta = 0.172$). Its interference colors are weak pastels of such high order that they are not visible. In contrast, Plates 13d and 13f show twinned plagioclase ($\delta = 0.011$) with first-order gray and white interference colors. If there is some ambiguity, first-order white and high-order white can be distinguished by inserting a full wave accessory plate. The plate will change first-order white to yellow or blue but will have little effect on high-order white. Most minerals show interference colors between those of plagioclase and calcite.

Table 3 (Part II) and Appendix E list minerals in order of their interference colors, often a key property for identifying minerals in thin section. Minerals with very low birefringence that display first-order white, gray, or yellow interference colors in thin section include leucite, nepheline, apatite, beryl, quartz, and feldspar. At the other extreme, minerals such as titanite (sphene), calcite, dolomite, and rutile display extreme retardation; interference colors are light pastels of high order. They may have such weak colors that it is hard to determine the retardation and birefringence with certainty.

Anisotropic minerals have different refractive indices depending on the path light travels when passing through them. Their optical properties, including birefringence, and thus interference colors, depend on their orientation. For identification purposes, the maximum birefringence (δ), corresponding to the highest-order interference colors, is diagnostic. This may be hard to estimate in grain mounts because mineral thickness varies, making it difficult or impossible to estimate birefringence from interference colors. In thin sections, the task is easier because thickness is known (0.03 mm), so we can use the Michel-Lévy Chart (Plate 2) to determine birefringence from interference color. Randomly oriented mineral grains may not show maximum interference colors; often we must rotate the stage and look at many grains of the same mineral. Because it is difficult to be exact, we normally use terms such as "low," "moderate," "high," or "extreme" to describe retardation and birefringence.

Some minerals have *anomalous interference colors,* colors that are not represented on the Michel-Lévy Color Chart. Anomalous interference colors may result if minerals have highly abnormal dispersion, if they are deeply colored, or for a number of other reasons. Minerals that commonly display anomalous inference colors include chlorite, epidote, zoisite, jadeite, tourmaline, and sodic amphiboles (see list in Table 3, Part II). Plates 26d and 26f show anomalous blue interference colors of one variety of chlorite.

▶ UNIAXIAL AND BIAXIAL MINERALS

Isotropic minerals have the same light velocity and therefore the same refractive index (n) in all directions. This is not true for anisotropic minerals, whether they are uniaxial or biaxial. For uniaxial minerals, we need two indices of refraction (ε and ω) to describe the mineral's refractive index. For biaxial minerals, we need three (α, β, and γ).

Optic axes are directions that light can travel through a crystal without being split into two rays (Figure 15). In some *uniaxial minerals,* the optic axis is parallel or perpendicular to crystal faces; in *biaxial minerals,* the two optic axes rarely are. Light traveling parallel to the single optic axis of a uniaxial mineral travels as an ordinary ray and has a unique refractive index, designated ω. Light traveling in other directions is doubly refracted, splitting into two rays with one having refractive index ω. The other has refractive index ε', which varies depending on the direction of travel. ε' may have any value between ω and ε, a limiting value corresponding to light traveling perpendicular to the optic axis. If $\omega < \varepsilon$, the mineral is uniaxial positive (+). If $\omega > \varepsilon$, the mineral is uniaxial negative (−). We sometimes use the mnemonics POLE (positive = omega less than epsilon) and NOME (negative = omega more than epsilon) to remember these relationships. The maximum possible value of birefringence in uniaxial crystals, δ, is $|\omega - \varepsilon|$ (Table 1). We can only see maximum birefringence if the optic axis is parallel to the microscope stage.

Most minerals are biaxial, having two optic axes (Figures 15c and 15d). Light passing through a *biaxial crystal* experiences double refraction unless it travels parallel to an optic axis. We describe the optical properties of biaxial minerals in terms of three mutually perpendicular directions: X, Y, and Z (Figure 16). The vibration direction of the fastest possible ray is designated X, and that of the slowest is designated Z. The indices of refraction for light vibrating parallel to X, Y, and Z are α, β, and γ. α is therefore the lowest refractive index, and γ is the highest. β, having an intermediate value, is the refractive index of light vibrating perpendicular to an optic axis.

Normally, light passing through a randomly oriented biaxial crystal is split into two rays, neither of which is constrained to vibrate parallel to X, Y, or Z, so their refractive indices will be some values between α and γ. However, if the light is traveling parallel to Y,

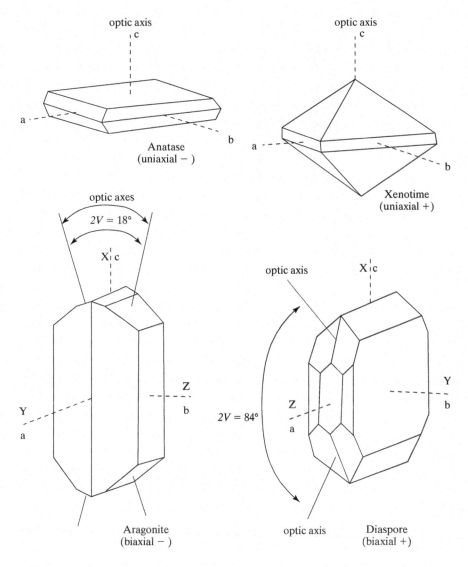

▶**FIGURE 15**
Sketches of crystals showing orientation of optic axes. (a) Anatase, TiO_2, uniaxial (−). (b) Xenotime, YPO_4, uniaxial (+). (c) Aragonite, $CaCO_3$, biaxial (−). (d) Diaspore, AlO(OH), biaxial (+)

TABLE 1

Indices of Refraction and Birefringence for Light Passing through Isotropic and Anisotropic Minerals

	Principal Indices of Refraction	Index of Refraction for Light Traveling Parallel to an Optic Axis	Indices of Refraction in a Random Direction	Birefringence in a Random Direction	Maximum Possible Birefringence
Isotropic crystals	n	n	n	0	0
Uniaxial crystals	ω, ε	ω	ω, ε'	$\delta' = |\omega - \varepsilon'|$	$\delta = |\omega - \varepsilon|$
Biaxial crystals	α, β, γ	β	α', γ'	$\delta' = \gamma' - \alpha'$	$\delta = \gamma - \alpha$

the two rays have refractive indices equal to α and γ, vibrate parallel to X and Z, and the crystal will display maximum retardation. If the light travels parallel to an optic axis, no double refraction occurs, and it has a single refractive index, β. There is no birefringence or retardation, and the mineral appears extinct.

In biaxial minerals, we call the plane that contains X, Z, and the two optic axes the *optic plane* (Figure 16). The acute angle between the optic axes is *2V*. A

line bisecting the acute angle must parallel either Z (in *biaxial positive* crystals) or X (in *biaxial negative* crystals). In biaxial positive minerals, the intermediate refractive index β is closer in value to α than to γ. In biaxial negative minerals, it is closer in value to γ. Retardation and *apparent birefringence* vary with the direction light travels through a crystal, but the maximum possible value of birefringence (δ) in biaxial crystals is always $\gamma - \alpha$.

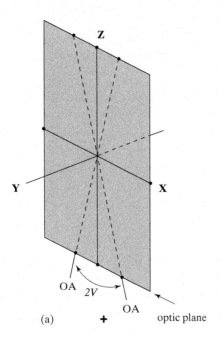

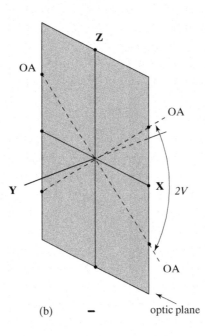

(a)   **+**   optic plane   (b)   **−**   optic plane

▶**FIGURE 16**
Geometric relationships between X, Y, Z, optic plane, optic axes (OA) and *2V* in biaxial positive and negative crystals. (a) In positive crystals, the acute angle between the optic axes is bisected by Z. (b) In negative crystals, the acute angle between the optic axes is bisected by X.

▶ACCESSORY PLATES AND THE SIGN OF ELONGATION

Polarizing microscopes have *accessory plates* that we can insert above the objective lens. When inserted, the slow and fast vibration directions of the plate are at 45° to the lower and upper polarizers. A double-headed arrow on accessory plates usually marks the slow direction. A standard full wave plate has a retardation of 550 nm (equal to the average wavelength or visible light), equivalent to first order red interference colors. A quartz wedge is sometimes a useful alternative to a full wave plate. The wedge has a variable thickness, with retardation ranging from 0 nm to 3,500 nm.

By inserting a plate when we are viewing a crystal on the microscope stage, we can add or subtract to the retardation. Accessory plates make it possible to learn which vibration direction in the crystal permits polarized light to travel the fastest. If crystals have a long dimension, we can learn whether the mineral is *"length fast"* (also sometimes called "negative elongation") or *"length slow"* ("positive elongation"). Determining the sign of elongation is often straightforward and can be helpful when identifying a mineral (Box 3).

▶UNIAXIAL INTERFERENCE FIGURES

Optic sign (positive or negative) is another useful characteristic for identifying anisotropic minerals. The easiest way to learn whether a uniaxial mineral is positive or negative is to examine an interference figure. Examples are shown in Plates 3–6. We obtain interference figures by passing conoscopic light through a mineral (Box 4). The *conoscopic lens* focuses light into the crystal from many different converging directions. After the light leaves the crystal and passes through the upper polarizer, we can insert a *Bertrand lens* to refocus the rays and magnify the interference figure.

To determine the optic sign of a uniaxial mineral, it is best to look down, or nearly down, the optic axis (Box 5). The figure obtained is an *optic axis figure* (OA figure). Examples of OA figures, also called uniaxial crosses, are shown in Plate 3. Finding grains that give an OA figure is normally not difficult. Grains oriented with the optic axis vertical appear isotropic because they have no retardation when being viewed down an optic axis. Grains oriented with the optic axis close to vertical have low birefringence and, therefore, low-order interference colors. An ideal uniaxial OA figure has a black cross that does not move much when the stage rotates (Figure 19a–c; Plate 3). But, even if the cross is somewhat off-center, we can use it to determine the optic sign. The center of the cross, called the *melatope*, corresponds to the direction of emergence of the optic axis. We call the dark bands forming the cross *isogyres*. The surrounding colored rings, if present, are *isochromes* (Plate 3c). They are bands of equal retardation caused by the light entering the crystal at slightly different angles. Minerals with low birefringence, like quartz, may not show isochromes (Plate 3a). When viewing an OA figure, we use an accessory plate to learn whether a uniaxial mineral is positive or negative (Box 5).

Box 3. Determining the Extinction Angle and the Sign of Elongation

Viewed with crossed polars, anisotropic grains go extinct every 90° as we rotate the microscope stage. We can measure the *extinction angle,* the angle between a principal cleavage or direction of elongation and extinction (Figure 17). Minerals with cleavages that exhibit *parallel extinction* go extinct when their cleavages or directions of elongation are parallel to the upper or lower polarizer (Figure 17a). Many monoclinic and all triclinic crystals exhibit *inclined extinction* and go extinct when their cleavages or directions of elongation are at angles to the upper and lower polarizer (Figure 17b). Some minerals exhibit *symmetrical extinction;* they go extinct at angles symmetrical with respect to cleavages or crystal faces (Figure 17c). Because an extinction angle depends on grain orientation, determining an extinction angle for minerals in thin sections requires measurements on a number of different grains, or on one grain in the correct orientation (determined by looking at interference figures, discussed later). If the grains are randomly oriented in the thin section and if the sample size is large enough, the maximum value of the measurements should approximate the actual extinction angle of the mineral.

Some anisotropic crystals have a prismatic habit or a well-developed cleavage that causes them to break into elongated fragments. Polarized light passing through an anisotropic prismatic crystal with polarization parallel to the long dimension will not

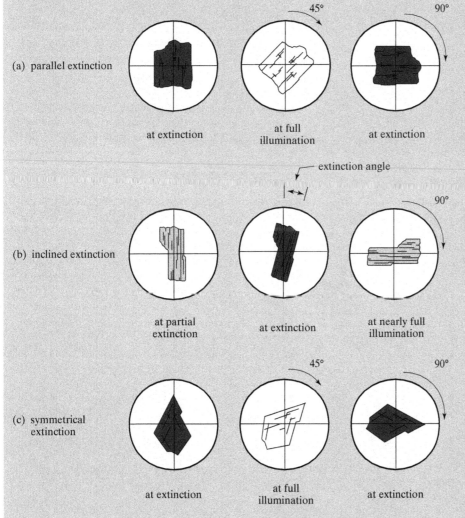

▶**FIGURE 17**
Extinction of minerals viewed with XP light in thin section. (a) Parallel extinction of minerals such as orthopyroxene occurs every 90° when the long axis of grains or the cleavage are perpendicular to either polarizer. (b) Inclined extinction in triclinic and many monoclinic minerals, including most micas, occurs when the cleavage is at an angle to both polarizers. (c) Symmetrical extinction of minerals, such as calcite, occurs when cleavages are oriented symmetrically to cross hairs and polarizers.

(a) parallel extinction

at extinction — at full illumination — at extinction

(b) inclined extinction

at partial extinction — at extinction — at nearly full illumination

(c) symmetrical extinction

at extinction — at full illumination — at extinction

Box 3. *Continued*

travel at the same velocity as light polarized in other directions. This distinction allows the prismatic minerals to be divided into two *signs of elongation:* "*length fast*" (faster light vibrates parallel to the long dimension) and "*length slow*" (slower light vibrates parallel to the long dimension). The sign of elongation cannot be determined on anisotropic crystals that cleave to produce equidimensional fragments.

Determining the sign of elongation (length fast or length slow) is usually straightforward for tetragonal and hexagonal prismatic crystals (all of which are uniaxial). We orient the long dimension of a grain in a southwest-northeast direction (45° to the lower polarizer) and note the interference colors. Many minerals, especially if grains are small, exhibit low first order interference colors (grays). We can insert a full wave accessory plate (having a retardation of 550 nm, equivalent to first-order red interference colors). After insertion, the slow direction of most accessory plates will be oriented southwest-northeast. If gray interference colors are added to first-order red, first-order blue results. If gray interference colors are subtracted from first-order red, first-order yellow results. For a grain oriented in the southwest-northeast direction, addition or subtraction is often just a matter of looking for blue or yellow. If we see blue or other higher order interference colors when the plate is inserted, the mineral is length slow. If we see yellow or other lower order colors, it is length fast (Figure 18).

If the interference colors for a mineral grain are not mostly gray, determining addition or subtraction may not be quite so simple. Sometimes it

will be necessary to rotate the stage 90° to see the colors that appear when the mineral is oriented northwest-southeast. The effects in that orientation will be opposite to those seen when the mineral is oriented southwest-northeast. When oriented northwest-southeast, higher order colors correspond to length fast, lower order to length slow. A quartz wedge can be useful in determining the sign of elongation if a grain contains several color bands rather than just gray. As the wedge is inserted into the accessory slot, the color bands on a southwest-northeast-oriented grain will move towards the thicker portions of the grain (usually the center of the grain) if the retardation is being subtracted (length fast). If the bands move away from the thicker portions of the grain, the retardation is being added (length slow).

Determining the sign of elongation for an orthorhombic, monoclinic, or triclinic mineral can be problematic or impossible. We can sometimes determine it for orthorhombic or monoclinic minerals with parallel extinctions, but it may vary with the orientation of the mineral. Mineral identification tables, such as those in Part II, sometimes list whether a given mineral is likely to provide a sign of elongation and whether the sign may vary with the mineral's orientation. If the extinction angle of the monoclinic or triclinic grain is only a few degrees, we can often determine a sign of elongation. If they have inclined extinction, we often cannot because the sign may vary in a complicated way with the orientation of the grain, and because determining the orientation of the grain on the microscope stage is difficult.

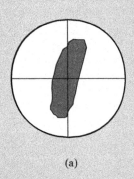
(a)

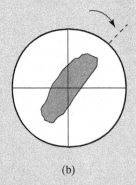
(b)

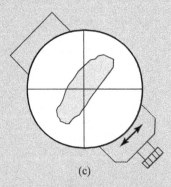
(c)

▶**FIGURE 18**
Determining the sign of elongation. (a) An elongated grain at extinction position (XP light). (b) The grain is rotated until it is at 45° to the cross hairs and oriented southwest-northeast. (c) The accessory plate (slow direction marked with a double arrow) is inserted to determine whether the interference colors add or subtract. If they add, producing blue or other higher order interference colors, the grain is "length slow"; if they subtract, producing yellow or other lower order interference colors, the grain is "length fast."

Box 4. Obtaining an Interference Figure

Any uniaxial or biaxial mineral (whether in a grain mount or a thin section) will, in principle, produce a visible interference figure; isotropic minerals will not. Care must be taken to choose grains without cracks or other flaws so that light can pass through without disruption. In addition, for some purposes it is necessary to find grains with a specific orientation.

Having chosen an appropriate grain, obtaining an interference figure is relatively straightforward. Carefully focus the microscope using PP light and high magnification. (If perfect focus is ambiguous, it may help to focus first at low magnification.) Insert the upper polarizer to get XP light and, if the microscope is properly aligned, the grain will still be in focus. Fully open the substage diaphragm, insert the substage conoscopic lens, and then insert the Bertrand lens above the upper polarizer. You will now see an interference figure, perhaps similar to one of those shown in Plates 3–6.

The conoscopic lens focuses light so that it enters the crystal from many different angles simultaneously. The Bertrand lens focuses the light so that it is parallel again when it reaches the eyepiece. In effect, the two lenses together permit the examination of light traveling through a crystal in many different directions. Without the lenses, we would have to look at many different crystals of known optical orientation to obtain the same information. Some older microscopes do not have Bertrand lenses, but interference figures can still be obtained by removing an ocular and inserting a peep sight, or by just peering down the tube. The figures, however, will be quite small.

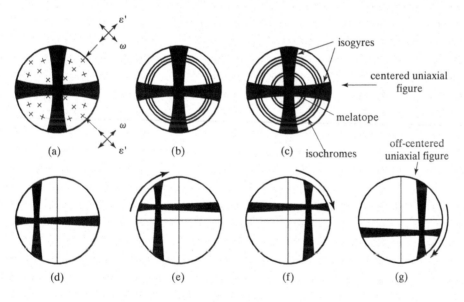

▶**FIGURE 19**
Uniaxial interference figures. In (a), (b), and (c) the figure is centered; in (d), (e), (f), and (g) it is off-center but the cross is visible. In (a), the small crosses show the vibration directions of ε' and ω; ε' always vibrates radially and ω vibrates perpendicular to ε'. In (b), one set of isochromes is shown, while in (c) there are two sets because of higher birefringence. (d) through (g) show how the melatope (the center of the cross) of an off-center figure precesses as the microscope stage is rotated.

Most uniaxial mineral grains do not exhibit a perfectly centered OA figure when viewed in conoscopic light. The optic axis is only one direction in the crystal and grains are unlikely to be oriented with the optic axis vertical. If we choose a random grain, we typically get an off-center figure (Figure 19d–g). If the optic axis of a grain lies parallel to the stage of the microscope, we get an *optic normal figure,* also called a *flash figure.* The flash figure appears as a vague cross or blob that nearly fills the field of view when the grain is at extinction (when the optic axis is perpendicular to one of the polarizers). Upon stage rotation, it splits into two curved isogyres flashing in and out of the field of view with a few degrees of stage rotation. Plate 6d–g shows a diffuse biaxial obtuse bisectric figure (Bxo). Uniaxial flash figures have the same general appearance.

Box 5. Determining the Optic Sign of a Uniaxial Mineral

To determine the optic sign of a uniaxial mineral, it is necessary to know whether $\omega > \varepsilon$ or $\varepsilon > \omega$. We do this by examining an optic axis figure and using an accessory plate with known orientation of the fast and slow rays. Standard accessory plates have their slow direction oriented southwest-northeast at 45° to both polarizers. The plate is inserted, and we observe interference color changes or isochrome (color ring) movements in the southwest and northeast quadrants of the interference figure (Figure 20).

Uniaxial optic axis figures appear as black crosses (Figures 19, 20; Plates 3a, 3c). In all parts of a uni-

axial optic axis interference figure ε' vibrates radially (along the radius of the interference figure) and ω vibrates tangent to isochromes (see Figure 19a). One way to remember these relationships is with the mnemonic WITTI (ω is tangent to isochrome). If interference colors in the southwest and northeast quadrants shift to higher orders when the plate (or wedge) is inserted, addition of retardation has occurred. ε' is the slow ray and the crystal is uniaxial positive. Subtraction (lower order interference colors) in the southwest and northeast quadrants indicates that ε' is the fast ray, and the crystal is optically

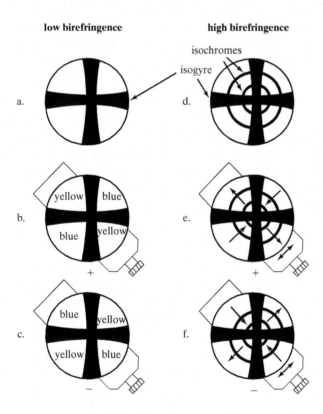

▶FIGURE 20

Determining the sign of a uniaxial mineral. (a) A slightly off-center uniaxial interference figure showing no isochromes (low birefrigence). (b and c) When the accessory plate is inserted, yellow and blue colors seen in different quadrants allow us to determine whether the mineral is positive or negative. (d) A slightly off-center uniaxial interference figure with isochromes (due to high birefrigence). (e and f) When a quartz wedge is inserted the isochromes move in and out in different quadrants depending on optic sign.

BIAXIAL INTERFERENCE FIGURES

We obtain *biaxial interference figures* in the same way as uniaxial interference figures. However, complications arise with biaxial minerals because it is more difficult to find and identify grains oriented in a useful way. We can get interference figures from all grains, but interpreting them can be difficult or impossible. Three types of interference figures are commonly identified (Box 6). When we observe an acute bisectric figure (Bxa) for a grain in an extinction orientation (Y perpendicular to one of the polarizers), it appears as a black cross, similar in some respects to a uniaxial interference figure. When we rotate the stage, the cross splits into two isogyres that move apart and may leave the field of view (Figure 21; Plate 5a–d). After a rotation of 45°, the isogyres are at maximum separation; they come back together to reform the cross with further stage rotation. The maximum amount of isogyre separation depends on *2V* (Box 7). If *2V* is less than about 60°, the isogyres stay in the field of view as we rotate the stage. If *2V* is greater than 60°, the isogyres completely leave the field of view. Plates 5a–h show the Bxa figures for topaz ($2V \simeq 60°$ and muscovite ($2V \simeq 40°$).

The points on the isogyres closest to the center of a Bxa or Bxo, the *melatopes,* are points corresponding to the orientations of the optic axes (Figure 24). If the retardation of the crystal is great, *isochromes* circle the melatopes. Interference colors increase in order moving away from the melatopes because retardation is greater as the angle to the optic axes increases.

The isogyres in a Bxo figure always leave the field of view because, by definition, an obtuse angle sepa-

Box 5. Continued

negative. Because ε′ vibrates parallel to the prism axis (c-axis) in prismatic uniaxial minerals, the optic sign is the same as the sign of elongation.

Although we could use a quartz wedge, we normally use a full wave accessory plate when the optic axis figure shows low order interference colors. In positive crystals, addition produces blue in the southwest and northeast quadrants of the interference figure, while subtraction produces yellow in the northwest and southeast quadrants (Plates 3; Figure 20). In negative crystals, the effect is the opposite. If several different color rings (isochromes) are visible, the blue and yellow colors will only appear on the innermost rings near the melatope (the center of the black cross; see Plate 3c and 3d).

We can sometimes use a full wave plate, but normally use a quartz wedge, to determine an optic sign for minerals that exhibit high order interference colors. The retardation of the quartz will add to the retardation of the unknown mineral in two quadrants as the wedge is inserted. It will also subtract from the retardation in the other two. The result will be color rings (isochromes) moving inward in quadrants where addition occurs, and outward in quadrants where subtraction occurs (Figure 20). For positive minerals, this means that colors move inward in the southwest and northeast quadrants and outward in the northwest and southeast quadrants. For negative minerals, the motion is opposite.

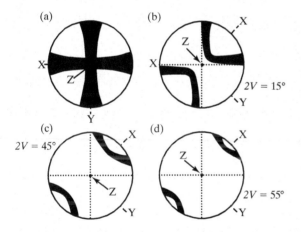

▶FIGURE 21
Biaxial acute bisectrix (Bxa) interference figures for a biaxial (+) crystal. In (a) the crystal is oriented with X and Y parallel to the polarizers; in (b), (c), and (d) the stage has been rotated 45° so that X (the trace of the optic plane) is oriented southwest-northeast and the isogyres are at maximum separation. The degree of separation is proportional to *2V*. Similar relationships apply to biaxial (−) crystals except that X and Z are interchanged. Compare these drawings with Plate 5a–h.

rates the optic axes in the Bxo direction (Plate 6d–g). Thus, if the isogyres remain in view, the figure is a Bxa figure. For standard lenses, if they leave the field of view, the figure may be a Bxo or a Bxa for a mineral with high *2V* (greater than about 60°). Getting a per-

fectly centered Bxo or Bxa figure is difficult. It may only be possible to see one isogyre clearly. Experienced mineralogists can tell a Bxa from a Bxo figure by the speed with which the isogyre leaves the field of view on rotation. For the rest of us, it is probably best to search for another grain with a better orientation.

In the discussion of interference figures, we have assumed that the objective lens being used has a numerical aperture (N.A.) of 0.85, because this is a common N.A. value today. If you use a lens with a different N.A., some of the 2V angular values given may be in error.

Determining optic sign from a Bxa figure is equivalent to asking whether the Bxa corresponds to the fast direction (biaxial negative crystals) or to the slow direction (biaxial positive crystals). We can make the determination in much the same manner as for a uniaxial interference figure (Box 7). However, for figures where the isogyres leave the field of view, this is not the recommended way to learn optic sign, because Bxa and Bxo figures are so easily confused when *2V* is greater than 70° to 80°.

An alternative way to determine optic sign and *2V* is to find a grain that yields a centered optic axis figure (Plate 4a–d). Finding such grains is often not difficult because such grains have little or no retardation (Box 8). If we look directly down an optic axis, a grain appears isotropic. We can learn the optic sign by obtaining an interference figure and using the full wave plate, and we can estimate *2V* by the curvature of the isogyre (Box 8).

Box 6. The Four Kinds of Oriented Biaxial Interference Figures

Biaxial interference figures are more difficult to interpret than uniaxial interference figures. With care, however, it is possible to identify three different kinds of useful figures: optic axes, acute bisectrix, and optic normal. Each corresponds to a different light path through the crystal, relative to the orientation of the optic axes (Figure 22). A fourth type of figure (Bxo) is described below for completeness, but Bxo figures are generally not useful.

Optic axis figure (OA): We obtain an OA figure by looking down an optic axis (Figures 22 and 23). We can locate grains oriented to give an OA figure because they have zero or extremely low retardation. Under orthoscopic XP light, they remain dark even if the microscope stage is rotated. An interference figure will show only one isogyre unless *2V* is quite small (less than 30°). We can use the curvature of the isogyre to estimate *2V* (see Box 8; Figure 23).

Acute bisectrix figure (Bxa): We obtain Bxa figures, such as those in Figures 21 and 24 and Plates 5a–h, by looking down the *acute bisectrix*, the line bisecting the acute angle between the two optic axes (Figure 22). In biaxial positive crystals, Bxa corresponds to a view along Z; in biaxial negative crystals, it corresponds to a view along X. For small values of *2V*, the isogyres form a black cross similar to a uniaxial interference figure when Y is parallel to the lower polarizer or upper polarizer. On rotation of the stage, the isogyres will split into hyperbolas, reforming the cross every 90° (Figure 24; Plates 5a–h). For *2V* values greater than about 60°, the isogyres will completely leave the field of view before coming back together to reform the cross.

Optic normal figure: An ON is the figure obtained when looking down Y, normal to the plane of the two optic axes (Figure 22). Grains that yield an optic normal figure are those that have maximum retardation. The interference figure resembles a poorly resolved Bxa, but the isogyres leave the field of view with only a slight rotation of the stage. Biaxial optic normal figures appear similar to uniaxial flash figures.

Obtuse bisectrix figure (Bxo): Bxo figures are generally not useful for mineral identification, but they can be confused with other, more useful figures. A Bxo is the figure obtained when looking down the *obtuse bisectrix*, a line bisecting the obtuse angle between the optic axes (Figure 22). A Bxo looks superficially like a Bxa, but the isogyres will always leave the field of view on a stage rotation (because the angle between the optic axes is greater than 90° along the Bxo; see Plate 6d–g). Distinguishing a Bxo from a Bxa is difficult or impossible for large values of *2V* (greater than 70°). Bxo figures are also easily confused with ON figures in minerals with small *2V*.

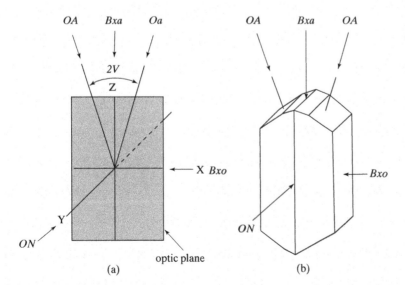

▶**FIGURE 22**
Relative orientations of optic axis (OA), Bxa, Bxo, and optic normal (ON) directions in a biaxial (+) crystal. (a) Schematic showing optic axes and optic plane. (b) Drawing of a crystal in a similar orientation. In biaxial (−) crystals the relationships are similar, but the X and Z axes are interchanged.

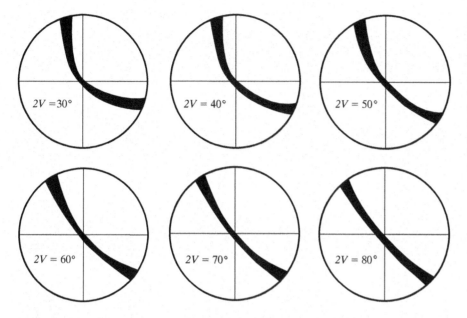

▶**FIGURE 23**
Biaxial optic axis (OA) figure. These figures show the isogyres that would be seen when looking down an optic axis for crystals with various *2V* values. The microscope stage has been rotated so the isogyre is concave to the northeast. The curvature can be used to estimate *2V*. Compare these drawings with Plate 4a–d.

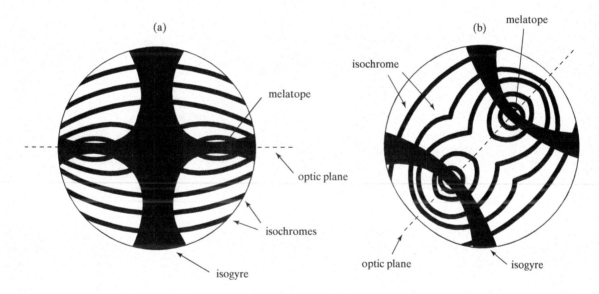

▶**FIGURE 24**
Two views of a Bxa interference figure. (a) A black cross, with or without isochromes, is seen when the optic plane is parallel to a polarizer. (b) The cross separates into two curved isogyres when the optic plane is parallel to neither polarizer. The optic plane is oriented northeast-southwest. Compare these figures with Plate 5a–h.

Box 7. Determining Sign and 2V from a Bxa Figure

Finding a grain that yields a Bxa is easiest for minerals with small *2V*. You can begin by searching for a grain that has minimal retardation. Such a grain will be oriented with one of the optical axes near vertical. Then obtain an interference figure, rotate the stage, and note how the isogyres behave. After checking several grains, you should find a Bxa. For minerals with low to moderate *2V* (0 to 60°), both isogyres will stay in the field of view, but the interference figure may not be perfectly centered (Plate 5e–h). If one of the isogyres leaves the field of view, check other grains until you are sure you are looking at a nearly centered Bxa. For minerals with high *2V*, the search for a Bxa sometimes becomes frustrating. It may be difficult or impossible to distinguish a Bxa from a Bxo. For this reason, the optic sign often is best determined using an OA figure (Box 8).

Having found a Bxa, rotate the stage so that the isogyres are in the southwest and northeast quadrants (Figures 24b; 25a–b). The Y direction in the crystal is now oriented northwest-southeast. The points corresponding to the optic axes (the melatopes) are the points on the isogyres closest to each other, and either X or Z is vertical, depending on optic sign.

In order to determine optic sign, we must know which direction (X or Z) is vertical, corresponding to Bxa. If the slow direction (Z) corresponds to Bxa, the crystal is positive. If the fast direction (X) is Bxa, the crystal is negative. To make the determination, insert the full wave accessory plate (with slow oriented direction southwest-northeast) and note any changes in interference colors on the concave sides of the isogyres. If the interference colors add on the concave sides of the isogyres (and subtract on the convex sides), the mineral is positive (Figure 25a–b). In positive minerals with low to moderate retardation, the colors in the center of the figure will be yellow (subtraction), and those on the concave side of the isogyres will be blue (addition). In a negative mineral, the color changes will be the opposite.

For minerals with high retardation, it may be difficult to determine whether a full wave accessory plate adds or subtracts to retardation because isochromes of many repeating colors circle the melatopes. A quartz wedge facilitates determination. As the wedge is inserted, color rings move toward the melatopes if there is addition of retardation, or the rings move away from the melatopes if there is subtraction. If the interference colors move toward the optic axes from the concave side of the isogyres, and away on the convex side, the mineral is positive. The opposite effect is seen for a negative mineral.

You can estimate *2V* from a Bxa figure by noting the amount separation of the isogyres when you rotate the stage. For standard lenses, if the isogyres just leave the field of view when they are at maximum separation, *2V* is 60–65° (Plate 5a–d). If the isogyres barely separate, *2V* is greater than about 10°.

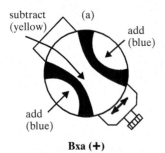
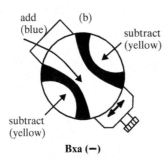
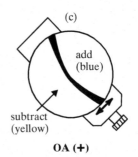
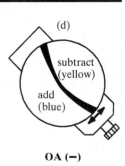

▶**FIGURE 25**

Determining optic sign from Bxa and OA figures. The accessory plate (with slow direction oriented southwest-northeast) is inserted and interference color changes are noted. For a biaxial positive mineral, interference colors add (increase to higher order) on the concave side of the isogyres; for a biaxial negative mineral, they subtract. For minerals with low retardation, addition produces blue interference colors and subtraction produces yellow. (a) a Bxa figure for a biaxial positive mineral; (b) a Bxa figure for a biaxial negative mineral; (c) an OA figure for a biaxial positive mineral; (d) an OA figure for a biaxial negative mineral. See Plates 4–5.

Box 8. Determining Sign and 2V from an Optic Axis Figure

Determining the optic sign from an optic axis figure can often be simpler than looking for a Bxa. To find an OA figure, look for a grain displaying zero or very low retardation. Obtain an interference figure and rotate the stage. It should be an OA figure, showing one centered or nearly centered isogyre (Figures 23 and 25c–d; Plate 4). If you see two isogyres that stay in the field of view when you rotate the stage, you are looking at a Bxa for a mineral with low *2V.* Rotate the stage so that the isogyre (or the most nearly centered if there are two) is concave to the northeast (Figure 25). Note that the isogyre rotates in the opposite sense from the stage. Insert the full wave accessory plate. If the retardation increases on the concave side of the isogyre (and decreases on the convex side), the mineral is positive (Figure 25c–d; Plate 4b). In minerals with low to moderate retardation, it is often necessary only to look for yellow and blue. Blue indicates an increase and yellow a decrease in retardation. The increase or decrease of retardation will be opposite if the mineral is negative (Figure 25d; Plate 4d).

You can estimate *2V* by noting the curvature of the isogyre. If *2V* is less than 10° to 15°, the isogyre will seem to make a 90° bend. If *2V* is 90°, it will be straight. For other values, it will have curvature between 90° and 0° (Figure 23).

Other Mineral Characteristics in Thin Sections

Besides the optical properties already discussed, we can use several other mineral characteristics to aid mineral identification. These characteristics give certain minerals a distinctive property in thin section. An experienced microscopist can, for example, often identify plagioclase feldspar because of its twinning. Other important characteristics include distinctive cleavage, alteration, compositional zonation, exsolution, anomalous extinction, or the presence of inclusions.

▶CLEAVAGE

Many minerals exhibit *cleavage;* when it can be seen with a microscope, it can be an important diagnostic tool. We use qualitative terms such as *perfect, good, fair,* and *poor* to describe the ease with which a mineral cleaves in different directions. Cleavage appears as fine parallel cracks in mineral grains when viewed with a microscope. Minerals with one or more good or perfect cleavages can be expected to show cleavage most of the time, while those with only poor cleavage may not. Additionally, minerals with low relief do not show cleavage as readily as those with high relief. This problem can be overcome somewhat by closing down the substage diaphragm, which narrows the cone of light hitting the thin section and increases contrast.

Minerals may have zero, one, two, three, four, or even more cleavages, but, because thin sections provide a view of only one plane through a mineral grain, we rarely see more than three at a time. Minerals that have elongate habits generally exhibit different cleavage patterns when viewed in a cross section than they do when viewed in a longitudinal section. Amphibole, for example, shows two good cleavages intersecting at

about 60° and 120° in a cross section, but only one good cleavage in a long section. The number of different cleavages and the angles between them aid mineral identification. Amphibole cross sections often show a typical diamond cleavage pattern (Plate 22c), which serves to distinguish amphiboles from other similar minerals. However, it is important to remember that cleavage angles depend on grain orientation. If a mineral has two cleavages at 60° to each other, the cleavages will appear to intersect at any angle from 0° to 60° depending on grain orientation. So we must often examine many grains (or one with a known orientation determined by examining an interference figure) to determine the maximum, and true, cleavage angle.

▶TWINNING

Many minerals *twin,* and sometimes we can see the twins with a microscope (see the feldspars in Plate 13, for example). They manifest themselves as different regions of a grain that have different crystallographic orientations, so they do not go extinct at the same time when the microscope stage is rotated. *Contact twin* domains are separated by a sharp line, the trace of the twin plane. *Penetration twins* generally have irregular domain boundaries. Simple twins consist of two individual *domains* (see, for example, the pyroxenes in Plates 20a–b), but *lamellar twins* are characterized by multiple parallel bands called *twin lamellae.* The plagioclase in Plate 13d shows well-developed twin lamellae. Some minerals, such as the plagioclase in Plate 13f, exhibit *polysynthetic twinning;* many parallel twin lamellae, often quite narrow, related by parallel twin planes. Still others, such as andalusite (Plate 29a–b) may exhibit *cyclic twinning,* although it is rarely seen in thin sections. The feldspars are excellent examples of minerals that twin. Plagioclase is characterized by polysynthetic twinning, orthoclase often by simple penetration or contact twins, and microcline (Plate 13b) by two types of lamellar twins with different orientations (that combine to produce "scotch plaid" twinning). Calcite is characterized by polysynthetic twins parallel to the long diagonal of its rhombohedral shape (Plate 8d). Other carbonates have no twins or have twins parallel to the short diagonal. Thus, for the feldspars, the carbonates, and for other minerals, twinning can be a key to identification.

▶ ALTERATION

Many minerals *alter* due to weathering, the circulation of hydrothermal fluids, retrograde metamorphism, or for other reasons (see Plates 19c, 26c, 26e, 28e, and 29a). Alteration products are often referred to as *secondary minerals.* Sometimes alteration results in the complete replacement of one mineral by another, leaving pseudomorphs (relict shapes) behind. Alteration can also result in partial replacement of one mineral by another. Chlorite, for example, can often be seen as a replacement for biotite or other ferromagnesian minerals in a thin section (Plate 26c). Sericite, a micaceous intergrowth, can appear to have grown over or to have replaced feldspar (Plate 14). Alteration is sometimes helpful for mineral identification. Pinite (a fine-grained greenish yellow mix of sericite and other minerals) normally indicates the former presence of cordierite. It usually appears along cracks and grain boundaries, sometimes leaving a core of ragged cordierite behind. Iddingsite, a fine-grained red or yellow-brown alteration product, is a characteristic alteration product of some olivine (Plate 19c).

▶ ZONING

Many mineral grains are compositionally *heterogeneous.* The heterogeneity can be primary (due to events that occurred at the time of initial crystallization), or secondary (a result of alteration or metamorphism). In thin sections, a heterogeneous crystal can appear *zoned* if areas of different composition have different optical properties. Plates 12a–d, and 14a–b show zoned feldspar crystals in igneous rocks. Different zones in the crystals have slightly different optical properties, so the feldspar shows concentric black-gray-white patterns when viewed with XP light.

Besides compositional variations, the presence of other minerals, or of trapped fluids, can cause minerals to appear zoned.

▶ EXSOLUTION

At high temperatures, some mineral compositions are stable that are not stable at low temperatures. Consequently, some minerals *exsolve* (unmix) to form two distinct compositions during or after cooling. If the unmixing is complete, separate grains of different compositions will form; it may be impossible to tell they were once one. Sometimes, however, *exsolution lamellae* form instead of discrete grains. Plate 16a–d show *perthite* grains composed of two different alkali feldspars: one is sodium-rich and the other is potassium-rich. Perthite forms when originally homogeneous high temperature alkali feldspar unmixes. In XP light, the grain has a gray and white striped appearance because the two feldspars have different orientations and properties. Perthite may exhibit a blebby, rather than a striped, appearance. Plate 34c shows another example of exsolution: an exsolved grain containing magnetite and ilmenite. The two oxides have slightly different colors and reflectivities. Besides feldspars and oxides, pyroxenes, amphiboles and other minerals commonly display exsolution in thin sections. Exsolution is sometimes confused with twinning, but exsolution lamellae are normally less regular, are not bounded by planar surfaces, and are of variable width.

▶ DISTINCTIVE EXTINCTION

We have already discussed parallel, inclined, and symmetrical extinction (Figure 17). These can be useful for identifying minerals showing an elongate habit or prominent cleavage. Two other distinctive types of extinction are *undulatory extinction* and *bird's eye extinction.* When different parts of a mineral grain go extinct at slightly different times during stage rotation, giving a wavy or blotchy appearance near the extinction position, we call it *undulatory extinction.* Undulatory extinction is due to strain (giving different parts of the grain slightly different orientations) or due to chemical heterogeneity, and is more common in some minerals than in others. It is an especially common property of quartz and helps distinguish quartz from feldspars and other minerals with low birefringence (Plate 12f).

Bird's eye extinction is a typical property of biotite and other micas. It helps distinguish biotite from brown amphiboles. It gives biotite, and some other micas, a pebbly or mottled appearance, sometimes described as a *bird's eye maple structure*. It is often most easily seen when a grain is near extinction (Plate 18d and 25f). We can also see bird's eye patterns in carbonates and a few other minerals.

▶INCLUSIONS

Minerals often contain inclusions of other minerals, or inclusions of trapped fluid, that affect their appearance in thin section. The presence or absence of inclusions can be a useful property for identification. Garnets, for example, typically contain inclusions of quartz and other minerals (Plate 1d). Staurolite, too, frequently contains inclusions that give it a *Swiss cheese* appearance (Plate 30c). Orthopyroxene sometimes contains small flake-like inclusions giving it a *schiller structure*. Cordierite can contain included sillimanite needles (Plate 31c and e), and some quartz contains rutile needles (Plate 18e–f). When inclusions are very small, they give minerals a cloudy appearance. Inclusions of uranium bearing minerals, such as zircon, can produce *pleochroic halos* (due to damage caused by radioactive decay; see Plates 18a and c and

32a). Pleochroic halos, appearing as dark circular patches or small burns that may be pleochroic, are especially common in biotite and some cordierite. The inclusion that contained the elements that produced the halo may or may not be visible.

▶OPAQUE MINERALS IN THIN SECTIONS

Opaque minerals (examples in Plates 33 and 34) can be found in most thin sections, but they may be difficult or impossible to identify. Table 3, Part II, lists the most common of them and gives their colors in reflected light. These colors can sometimes be seen by shining a light on the top of a thin section and observing the reflected color. This method usually serves to distinguish sulfides (yellow and yellowish minerals; see Plates 33f and 34b) from oxides or graphite (silverish minerals; see Plate 34c), and perhaps to distinguish leucoxene (white) or hematite (red; see Plate 34f). Telling the various sulfides apart, or distinguishing magnetite from ilmenite, can be impossible unless they are side by side. For proper opaque mineral identification, it is usually necessary to use a reflecting light microscope or some other instrument.

REFERENCES

Bambauer, H.U., F. Taborszky, and H.D. Trochim. *Optical Determination of Rock-Forming Minerals*. Stuttgart: E. Schweizerbart'sche Verlagsbuchhandlung, 1979.

Bloss, F.D. *An Introduction to the Methods of Optical Crystallography*. New York: Holt, Rinehart and Winston, 1961.

Deer, W.A., R.A. Howie, and J. Zussman. *Rock Forming Minerals*. Five vols. New York: John Wiley & Sons, 1962 et seq.

Gaines, R.V., H.C.W. Skinner, E.E. Foord, B. Mason, and A. Rosenzweig. *Dana's New Mineralogy,* Eighth ed. New York: John Wiley & Sons, 1997.

Kerr, P.F. *Optical Mineralogy,* Fourth ed. New York: McGraw Hill, 1977.

Klein, C., and C.S. Hurlbut, Jr. *Manual of Mineralogy,* Twenty-first ed. New York: John Wiley & Sons, 1993.

Nesse, W.D. *Introduction to Optical Mineralogy,* Second ed. New York: Oxford University Press, 1991.

Perkins, D. *Mineralogy*. Second ed. Upper Saddle River: Prentice Hall, 2001.

Phillips, W.R. *Mineral Optics: Principles and Techniques*. San Francisco: W.H. Freeman & Co., 1971.

Phillips, W.R., and D.T. Griffen. *Optical Mineralogy: The Nonopaque Minerals*. San Francisco: W.H. Freeman & Co., 1981.

Pichler, H., and C. Schmitt-Riegraf, translated by L. Hoke. *Rock-Forming Minerals in Thin Section*. Chapman and Hall, 1997.

Shelley, D. *Optical Mineralogy,* Second ed. New York: Elsevier, 1985.

Stoiber, R.E., and Morse, S.A. *Microscopic Identification of Crystals*. Krieger, 1979.

Zoltai, T., and J.H. Stout. *Mineralogy: Concepts and Principles*. Minneapolis: Burgess, 1984.

PLATE 1: DIFFERENCES IN RELIEF

Garnet, Beryl, Quartz, Biotite, opaques, Plagioclase

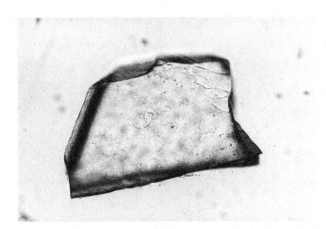
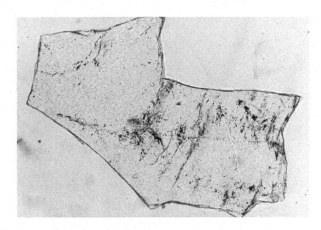

Photographs a and b: <u>Garnet</u> (high relief) and <u>Beryl</u> (moderate relief in oil), 100x magnification in plane-polarized (PP) light.

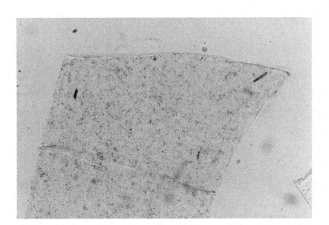
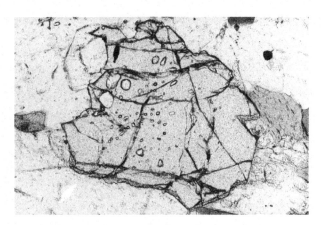

Photograph c: <u>Quartz</u> with very low relief in oil with about the same index of refraction. Notice the orange Becke line on the top of the grain (100x, PP). **Photograph d:** Relief comparisons in PP light (100x) in a thin section. A large light-pink <u>Garnet</u> (very high relief) is surrounded by light brown <u>Biotite</u> (moderate relief) and clear <u>Quartz</u> grains (relatively low relief). Because of the high index of refraction of <u>Garnet</u>, fractures and <u>Quartz</u> inclusions are clearly visible.

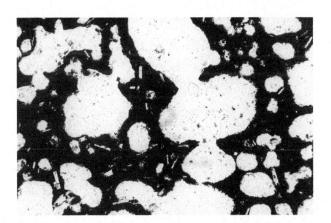
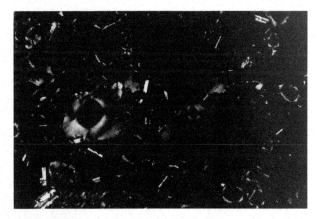

Photographs e: (PP, 40x) and **f (cross-polarized [XP] light, 40x) of the same area of a thin section:** Ward's collection, sample #34, a Scoria from Klamath Falls, Oregon. The thin section contains iron-rich opaques, <u>Plagioclase</u> laths, and holes from the vesicles. In XP light, the holes are usually black—but as shown in Photograph f, they may contain white halos that resemble a solar corona during an eclipse.

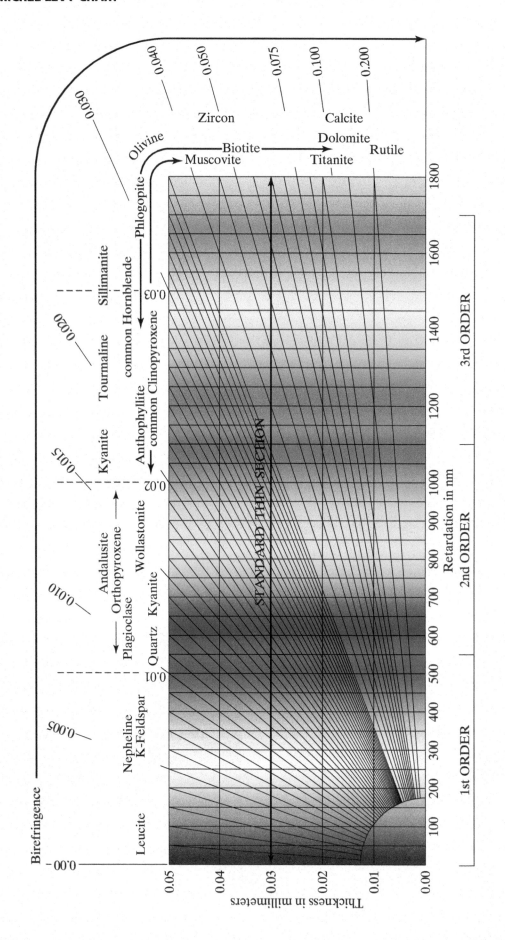

PLATE 3: UNIAXIAL OPTIC AXIS (OA) INTERFERENCE FIGURES

Quartz, Calcite

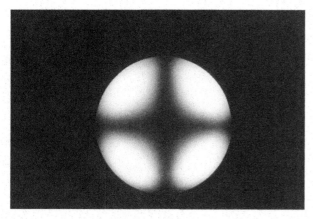

a

Photograph a: Quartz uniaxial cross (uniaxial optic axis interference figure).

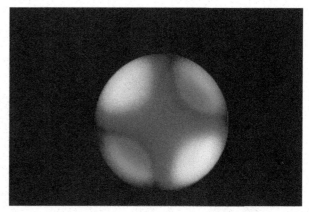

b

Photograph b: The same interference figure as in photograph a, with a 1-wavelength plate inserted. The blue in the northeast and southwest quadrants indicate that quartz has a positive (+) optic sign.

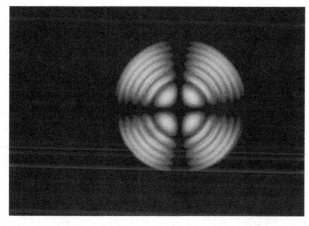

c

Photograph c: Calcite uniaxial cross (uniaxial optic axis interference figure) with abundant isochromes.

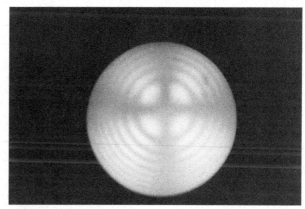

d

Photograph d: The same interference figure as in photograph c, with a 1-wavelength plate inserted. The yellow near the melatope in the northeast and southwest quadrants (and the blue in the northwest and southeast quadrants) indicate that calcite has a negative (−) optic sign.

PLATE 4: BIAXIAL OPTIC AXIS (OA) INTERFERENCE FIGURES

Augite, Epidote

a

Photograph a: Augite (biaxial) optic axis interference figure. The curvature of the isogyre indicates *2V* ≈ 60°.

b

Photograph b: The same interference figure as photograph a, with a 1-wavelength plate inserted. The blue in the northeast quadrant indicates a positive sign.

c

Photograph c: Epidote (biaxial) optic axis interference figure with many isochromes.

d

Photograph d: The same interference figure as photograph c, with a 1-wavelength plate inserted. Lower order interference figures in the northeast quadrant (including gray and yellow near the melatope) indicate a negative sign.

PLATE 5: BXA INTERFERENCE FIGURES

Topaz, Muscovite

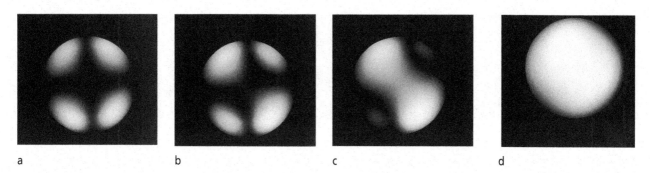

a   b   c   d

Photographs a–d: A series of photographs of a biaxial Bxa figure from topaz showing the separation of the isogyres as the microscope stage rotates. As the stage rotates, the cross in photograph a breaks into hyperbolic isogyres that begin to leave the field of view as shown in photographs b and c. A maximum separation (*2V*) of about 60° occurs and the isogyres are barely visible in the northeast and southwest quadrants of photograph d.

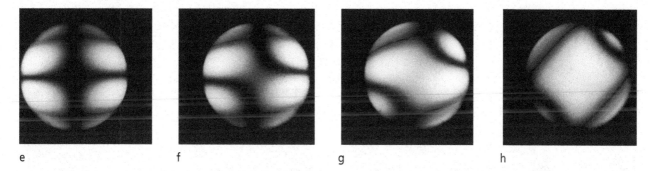

e   f   g   h

Photographs e–h: A series of photographs of a Bxa figure from muscovite showing the separation of the isogyres as the microscope stage rotates. The maximum separation (*2V*) is only 40° and the isogyres never leave the field of view.

PLATE 6: BXA INTERFERENCE FIGURES AND BXO INTERFERENCE FIGURES

Olivine, Barite

a

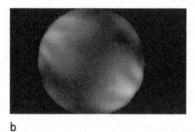

b

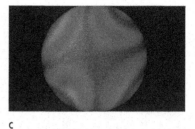

c

Photographs a–c: Olivine Bxa interference figures. In photograph a, the olivine is at extinction. In photograph b, the stage has been rotated 15° from extinction. In photograph c, the stage has been rotated 45° from extinction. For this olivine, *2V* is significantly greater than 60°. The isogyres leave the field of view.

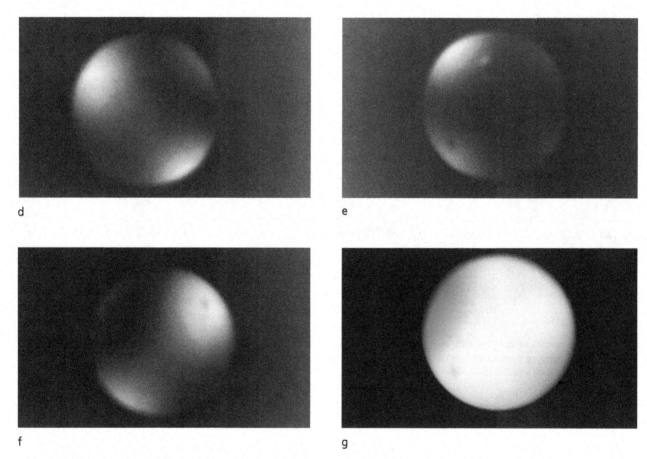

d

e

f

g

Photographs d–g: A series of photographs of a diffuse biaxial Bxo figure from barite showing the separation of the isogyres as the microscope stage rotates. Uniaxial flash figures have the same general appearances. In photograph d, two isogyres are merging from the northeast and southwest quadrants. The isogyres are slightly off-center with the northeast one extending slightly farther into the field of view. In photograph e, the isogyres form a diffuse cross. In photograph f, the isogyres are separating and leaving in the northwest and southeast quadrants. In photograph g, the isogyres have entirely left the field of view and are replaced by blue interference colors.

PART

II

Identifying Minerals in Thin Section

Mineralogists have described more than 3,000 minerals; most of which are rare. The following pages contain descriptions of about 60 minerals, including all the major rock-forming minerals you are likely to see in thin sections of common rocks. Some are abundant; others are widespread but exist in small amounts. Still others have been included because they have a special significance or special properties. If you are studying rocks of unusual composition or history, you may encounter minerals that are not listed here. If so, you should consult one of the optical mineralogy books listed at the end of Part I. Those books also will give you more details about specific varieties of some of the minerals included here.

The mineral descriptions start with mineral *name*, *formula*, and a list of any *photographs* in this book that display the mineral. The second entry describes the kinds of rocks in which the mineral is typically found (*occurrence*). The *distinguishing features* and *similar minerals* entries list the key optical properties used for mineral identification. The *properties* entry gives the mineral's crystal system and optical properties. Mineral properties have been tabulated in many different books; most of the property values given here were taken from Bambauer *et al.* (1979); a few were updated from more current sources. Because natural minerals vary slightly in composition and, therefore, properties, ranges are given for some values. Averages are given when variations are insignificant. Under *color*, we list the typical colors in (PP light) and the pleochroic formula. If a mineral is typically uncolored or extremely weakly colored, we have called it *colorless. Form* describes the numeral's typical shape. *Cleavage* describes the quality and number of cleavages that can be seen in thin sections. *Relief* refers to the way a mineral stands out in thin sections. It is a relative property that derives from the difference in a refractive index between the mineral and the mounting material. *Interference colors* describe the colors seen in XP light, observed for thin sections of normal thickness (0.03 mm). If sections are thinner or thicker, the colors will be different, and identifying minerals becomes problematic. If you are suspicious about mineral thickness, examine the interference colors of known minerals such as quartz or feldspars. Under *extinction and orientation*, we have listed some additional properties that aid mineral identification. Extinction and orientation are complicated for some minerals, especially monoclinic and triclinic ones, and we have omitted the complications here. Under *twinning*, we briefly describe the kinds of twins that may be present. In general, they are either *simple* or *polysynthetic*.

SYSTEMATIC IDENTIFICATION

Box 2 (see inside back cover of this book) gives a straightforward process that you can follow in order to determine a mineral's properties. In some cases, you may find that you have to go through all the steps in the process to identify a mineral. In others, you may find some diagnostic properties or other short cuts. An experienced microscopist knows how to focus on those properties that are most significant.

The tables on the following pages are intended to aid in identifying an unknown mineral. If the mineral is moderately to strongly colored, start by using Table 2. It lists all the common colors of the common minerals. If, instead, the mineral has only very weak colors or shows no color at all, start with Table 3. The first three parts of Table 3 list minerals that are opaque, isotropic, or show anomalous interference colors. For opaque minerals, Table 3 gives the color that the mineral appears in reflected light; for isotropic minerals, it gives the mineral's relief; for anomalous minerals, it gives the typical anomalous interference colors. The next four parts of Table 3 separately list minerals with very low birefringence (first-order gray and white interference colors), low birefringence (up to first-order yellow and red interference colors), moderate to high birefringence (about second- through fourth-order colors), and extreme birefringence (including pearl white). Within each list, minerals are separated by optic class (uniaxial or biaxial) and optic sign. Tables 2 and 3 can provide you with a list of possible minerals to check in more detail. However, you may occasionally encounter minerals that have anomalous properties not included in these tables, or perhaps not included in the detailed descriptions that follow. In such cases you will have to consult other, more detailed books.

The minerals and colors listed here are those most likely to be seen in thin sections. If a mineral shows only very weak colors, Table 2 may not be applicable. Additionally, there are other colors and minerals that have not been listed because they are very rare. Most of these minerals are pleochroic and so may change color with stage rotation.

TABLE 2
The Most Common Minerals That Show Colors in Thin Sections. (Bold type indicates minerals described in detail in this book. Italic font indicates rare coloration.)

Red	Pink Rose	Orange	Brown	Yellow	Green	Blue	Violet	Gray	Black
				Isotropic Minerals					
	garnet		**garnet** **sphalerite** **spinel**	**sphalerite**	**fluorite** **spinel**	**haüyne** **fluorite** **sodalite** **spinel**	**fluorite**		**spinel**
spinel *sphalerite*	*fluorite* *sodalite*		*fluorite* *glass*	*fluorite* *garnet* *sodalite* *spinel*	*garnet* *glass*			*glass* *sodalite*	*garnet*
				Anisotropic Minerals					
iddingsite piedmontite **rutile**	**andalusite** **orthopyroxene (hypersthene)** titanite		aegirine-augite **biotite** cassiterite chondrodite **collophane** **hornblende** iddingsite **phlogopite** **rutile** **siderite** **stilpnomelane** **tourmaline** **zircon**	**actinolite** allanite **biotite** **chloritoid** chondrodite **epidote** jarosite monazite **phlogopite** piedmontite **staurolite** **stilpnomelane** **titanite** **tourmaline** **zircon**	**actinolite** aegirine-augite **biotite** **chlorite** **chloritoid** crocidolite **epidote** **glauconite** **hornblende** **orthopyroxene** **riebeckite** **stilpnomelane** **tourmaline**	**chloritoid** aegirine crocidolite **glaucophane** **riebeckite** **tourmaline**	**corundum** piedmontite **tourmaline**	**apatite** **glaucophane**	**tourmaline**
allanite	*chlorite* *corundum* *piedmontite* *staurolite* *tourmaline*	*piedmontite* *staurolite*	*aegirine* *apatite* *monazite* *orthopyroxene* *titanite*	*cassiterite* *clinochlore* *glauconite* *glaucophane* *rutile*	*apatite* *rutile* *titanite (sphene)* *tourmaline*	*apatite* *cordierite* *epidote* *rutile*	*crocidolite* *rutile*	*cassiterite* *chloritoid* *chlorite* *glaucophane* *rutile* *titanite* *zircon*	*aegirine* *glauconite* *titanite*

TABLE 3
Minerals Described in this Book, Sorted by Opacity and Interference Colors

	Color in Reflected Light	Mineral	Page #
Opaque Minerals	Brass yellow	chalcopyrite	94
	Iron black to brownish-black	chromite	102
	Black	graphite	90
	Red, black, or steel blue	hematite	97
	Violet black to silver	ilmenite	100
	Steel blue black to silver	magnetite	101
	Brass yellow	pyrite	92
	Bronze to copper red	pyrrhotite	93

	Relief	Mineral	Page #
Isotropic Minerals	Low	sodalite	52
	Moderate	analcime	48
	Fairly high	fluorite	95
	High	spinel	99
	Very high	garnet	78
	Very high	sphalerite	91

	Typical Anomalous Colors	Mineral	Page #
Minerals that Sometimes Show Anomalous Interference Colors	blue	jadeite	70
	blue, yellow-brown	clinozoisite	89
	blue, brown, violet	chlorite	64
	subtle blue, green-yellow	epidote	89
	variable	garnet	78

Birefringence	Optic Class	Optic Sign	Mineral	Page #
Minerals with Very Low Birefringence (first-order gray and white)		+	leucite	50
			quartz	42
	Uniaxial		apatite	111
			beryl	54
		−	chabazite	51
			corundum	98
			nepheline	49
			scapolite	53
		+	heulandite	51
	Biaxial		antigorite	56
			chlorite	64
			clinozoisite	89
		−	kaolinite	57
			microcline	43
			orthoclase	43
			sanidine	43

Birefringence	Optic Class	Optic Sign	Mineral	Page #
Minerals with Low Birefringence (up to first-order yellow or red)		+	barite	109
			chloritoid	84
			clinozoisite	89
			gypsum	110
			natrolite	51
			riebeckite	76
			staurolite	83
	Biaxial	+ or −	chlorite	64
			chrysotile	56
			cordierite	55
			hypersthene	66
			jadeite	70
			orthopyroxene	66
			plagioclase	45
		−	andalusite	81
			epidote	89
			glaucophane	76
			vesuvianite	88
			kyanite	80
			stilbite	51
			wollastonite	71
Minerals with Moderate to High Birefringence (second- to about fourth-order)	Uniaxial	−	tourmaline	77
		+	anhydrite	108
			augite	68
			cummingtonite	73
			diopside	67
			gibbsite	103
			jadeite	70
			lawsonite	87
			olivine	79
			prehnite	65
			sillimanite	82
	Biaxial	−	actinolite	74
			anthophyllite	72
			biotite	60
			epidote	89
			grunerite	73
			hornblende	75
			lepidolite	62
			montmorillonite	57
			muscovite	61
			pyrophyllite	58
			talc	59
			tremolite	74
Minerals with Extreme Birefringence (fourth order and above; often high-order pearl white interference colors)		+	zircon	86
			siderite	106
	Uniaxial	−	calcite	104
			dolomite	107
			magnesite	105
	Biaxial	+	rutile	96
			titanite (sphene)	85

Detailed Mineral Description

▶**A. QUARTZ, FELDSPARS AND OTHER FRAMEWORK SILICATES**

Quartz and Chalcedony SiO_2

Plates

quartz: 1c–d, 3a–b, 7a–f, 8c–d, 9a–d, 10a–d, 12e–f, 16a–f, 17a–f, 18a–b, 21c–d, 23e–f, 25c–d, 26e–f, 27a–d, 29c–f, 30a–f, 31a–b, 31e–f, 34a–f
chalcedony: 7a–d, 9a–b
chert: 9e–f

Occurrence—Quartz is a common and essential mineral in many sedimentary, metamorphic and igneous rocks. Chalcedony, a microcrystalline variety of quartz, is most typically found as a filling or lining in cavities and cracks.

Distinguishing Features—Colorless, first-order white or weak yellow interference colors, lack of cleavage, lack of alteration, lack of twinning, and often shows undulatory extinction (Plate 12e–f). Rarely contains inclusions (Plate 18e–f) of other minerals.

Similar Minerals—Quartz is sometimes confused with apatite, cordierite, beryl, or scapolite, but none of these are uniaxial positive.

Properties and Interference Figure—Hexagonal; $\omega = 1.5442$, $\varepsilon = 1.5533$, $\delta = 0.0091$. Uniaxial (+) with no isochromes.

Color—Colorless in thin section.

Form—Typically anhedral but may be euhedral prismatic. Sometimes intergrown with orthoclase or microcline (creating a graphic granite), or with plagioclase (creating myrmekite, Plates 16e–f).

Cleavage—Normally absent, highly strained grains may show fracture.

Relief—Very low.

Interference Colors—Normally first-order whites and grays, rarely reaching first-order yellow.

Extinction and Orientation—A diagnostic property is wavy (undulatory) extinction (Plate 12e–f); ideally prismatic crystals show parallel extinction; basal sections are always extinct.

Twinning—Twinning is common but does not show in thin section.

K-feldspar (orthoclase, sanidine, microcline, perthite, anorthoclase) KAlSi₃O₈

K-feldspar (orthoclase, sanidine, microcline, perthite, anorthoclase) KAlSi_3O_8

Plates

orthoclase: 9c–d, 14c–f
microcline: 9c–d, 13a–b, 17e–f, 25c–d
perthite: 16a–d
anorthoclase: 15a–b

Occurrence—K-feldspars are common in many kinds of silicic igneous rocks. Orthoclase and microcline are also found in sedimentary rocks, such as arkoses, and in a variety of metamorphic rocks. Sanidine, the high temperature polymorph of orthoclase and microcline, occurs typically as phenocrysts in rocks such as trachyte or rhyolite. Anorthoclase is a high temperature feldspar having composition between sanidine and albite. Perthite, a K-feldspar-albite intergrowth, forms by exsolution of anorthoclase and other high temperature K-Na feldspars.

Distinguishing Features—K-feldspars are colorless, display first-order gray and white interference colors, and usually show one or two cleavages. Orthoclase and sanidine may show simple twins; microcline exhibits polysynthetic (Scotch plaid or spindle-shaped) twinning in some orientations. Microcline and orthoclase may include blob-like or lamellar intergrowths (called *perthite*; see Plate 16a–d) with albite.

Similar Minerals—If no cleavage or twinning show, K-feldspar may be confused with quartz, but quartz is usually clearer and is uniaxial; nepheline is uniaxial ($-$). Sanidine may be confused with quartz or orthoclase, but cleavage, twinning, or *2V* should distinguish it. Rarely, microcline may be confused with other feldspars; twinning usually serves to distinguish it.

Properties and Interference Figure—Orthoclase is monoclinic; biaxial ($-$); $2V = 60°$ to $65°$; $\alpha = 1.520$, $\beta = 1.525$, $\gamma = 1.527$, $\delta = 0.005$–0.007. Microcline is triclinic; biaxial ($-$); $2V = 77°$ to $84°$; $\alpha = 1.519$, $\beta = 1.523$, $\gamma = 1.525$, $\delta = 0.006$. Sanidine is monoclinic; biaxial ($-$); *2V* variable; $\alpha = 1.521$, $\beta = 1.526$, $\gamma = 1.527$, $\delta = 0.005$–0.007. Although biaxial ($-$), *2V* may be so small that it appears uniaxial. Good interference figures are hard to obtain for microcline, due to twinning. See page 44, Box 9.

Color—K-feldspar is normally colorless in thin sections but can be cloudy or pale brown due to inclusions or alteration.

Form—Orthoclase forms subhedral to anhedral crystals, phenocrysts, and spherulites. Sanidine most often occurs as phenocrysts. Microcline occurs as subhedral to anhedral crystals.

Cleavage—Orthoclase has one excellent, one good, and occasionally one imperfect cleavage. Sanidine and microcline have one perfect and one less perfect cleavage.

Relief—Low.

Interference Colors—Maximum interference colors are first-order gray and white.

Extinction and Orientation—Orthoclase's extinction angle varies from 0 to about 12° depending on orientation and composition; its principal cleavage is inclined at a small angle to the fast ray. Sanidine's extinction angle varies from about 0 to 5° depending on orientation. Microcline's extinction angle varies up to about 15° depending on orientation.

Twinning—Simple (Carlsbad) twins are common in orthoclase and sanidine. Several other types of simple twins also may be present. Albite and pericline twins are always present in microcline, giving wavy extinction and a Scotch plaid or spindly appearance in some grains. See drawings in Box 9.

Box 9. K-feldspar

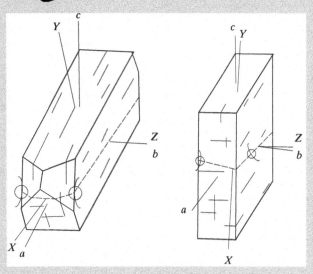

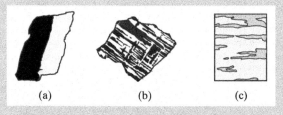

(a)   (b)   (c)

K-feldspar twinning and exsolution (XP): (a) simple twins are common and may correspond to one of several twin laws; (b) microcline (scotch plaid) twinning results from a combination of albite and pericline twinning; (c) exsolution may produce perthite with a blebby or wormy appearance.

K-feldspar crystals: Schematic drawings of orthoclase and low-temperature sanidine (left), and a cleavage fragment of microcline (right). The orientations of axes vary with composition and type of feldspar.

plagioclase

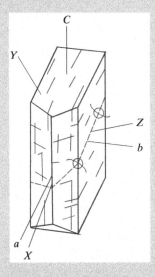

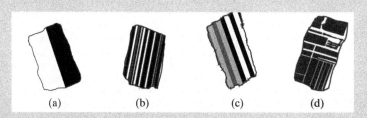

(a)   (b)   (c)   (d)

Plagioclase twinning: (a) carlsbad or simple twin; (b) polysynthetic twinning (albite or pericline) is most common; (c) albite and carlsbad twinning combined; (d) albite and pericline twinning combined.

Plagioclase crystal: orientation of axes and optic sign vary with composition.

Plagioclase   Solid solution of CaAl$_2$Si$_2$O$_8$ (anorthite) and NaAlSi$_3$O$_8$ (albite)

Plates—1e–f, 12a–d, 13a–f, 14a–b, 15e–f, 16a–f, 17a–f, 19c–f, 21c–d, 22a–d, 25c–f, 26c–f, 27c–d, 30a–b, 31a–b, 31e–f, 34a–b

Occurrence—Plagioclase feldspars are found in a wide variety of igneous and metamorphic rocks and, to a much lesser extent, in sedimentary rocks.

Distinguishing Features—Clear color, white to pale yellow interference colors, and polysynthetic (albite) twinning, giving it a black and white striped appearance (XP light, see Box 9, page 44); albite is sometimes inter-grown with microcline in perthite (often giving it a wormy black and white pattern under XP light); plagioclase may alter to sericite (fine grained white mica) or to sauserite (epidote). Zoning may be present (Plates 12b, 12d, 14b). Optical properties may be used to determine composition; see Box 10 (page 46).

Similar Minerals—May be confused with K-feldspars.

Properties and Interference Figure—Triclinic; biaxial (+ or −), 2V = 75–90°; optical properties vary with composition but refractive index (1.53 to 1.59) and birefringence (0.007 to 0.013) are low. See Box 9 (page 44).

Color—Colorless.

Form—Normally subhedral to anhedral plates or laths.

Cleavage—One perfect, one less perfect, and one poor.

Relief—Low.

Interference Colors—Maximum color is first-order light yellow.

Extinction and Orientation—Extinction angle varies with composition.

Twinning—Most plagioclase exhibits conspicuous albite twins (polysynthetic twins giving a striped appearance under XP light; see, for example, Plate 13d). Simple (Carlsbad) twins may also be present (Plate 12d), as may per-icline twins (Plate 13f). See Box 9 (page 44).

Box 10. Determining Plagioclase Composition: The Michel-Lévy Method

The Michel-Lévy method may be used to determine plagioclase composition (% anorthite). It is based on the observation that the angle of extinction from (010) composition planes of albite twins varies systematically with composition. To use this method, you must examine albite-twinned grains with (010) vertical, meaning perpendicular to the microscope stage (so you are looking in a direction parallel to the twin composition planes). Because grain orientation affects the observed extinction angle, you must examine several grains and use the largest extinction angle for determining composition. *Note: If you are examining plagioclase in volcanic rocks, you must use the dashed curves in the graph below.*

To identify grains with (010) vertical:

- View grains using XP light and select one where all twin lamellae have about the same interference color (shade of gray) when oriented north-south.
- Raise and lower the stage to ensure that the composition plane is vertical and remains sharp as focus changes.
- Rotate the stage to the right (until one set of twins goes extinct) and then to the left (so other set of twins goes extinct) and measure extinction angles L1° and R1° (drawing below). They should be equal or differ by less than 4°. For most plagioclase compositions, L1° and R1° correspond to the fast rays being parallel to the north-south polarizer. Continued rotation in either direction reveals a second ex-

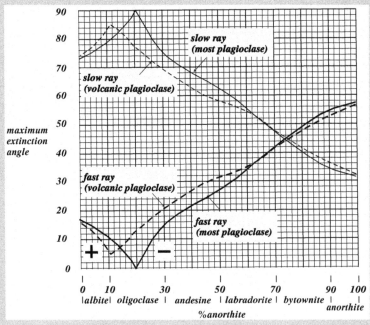

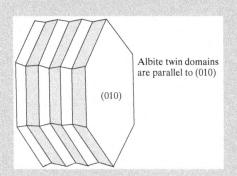

Albite twin domains are parallel to (010)

(010)

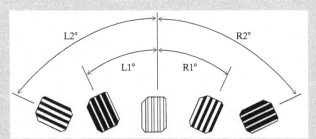

Box 10. Continued

tinction angle (L2° and R2°), occurring when the slow rays are oriented north-south. For very anorthite-rich plagioclase, this relationship is reversed.

To determine composition:

- Repeat the measurements on a number of grains (generally 6 or more) until confident you have determined the maximum value. If L1° and R1° are not equal, use their average value. If they differ by more than 4°, find a different grain. Once you have determined the maximum value, read the composition from the graph below.
- For maximum extinction values less than about 20°, there are two possible compositions. They may be told apart by determining optic sign. Plagioclase with composition An_{0-20} is optically positive while other plagioclase is optically negative.
- If maximum extinction angle is greater than 20°, you must determine if extinction angle you measured corresponds to the fast ray or the slow ray parallel to the north-south polarizer. To make this distinction, rotate the stage 45° clockwise from the extinction position and insert the accessory plate. If the interference colors in the lamellae decrease (usually to 1st order red or yellow), the fast ray extinction angle was measured. If the interference colors increase (usually to 2nd order blue), then the slow ray extinction angle was measured. In either case, composition can be read from the chart.

Analcime   **NaAlSi$_2$O$_6$•H$_2$O**

Occurrence—Analcime is found in cavities and cracks in basalt and as a primary mineral in alkalic igneous rocks such as Na-rich basalt or syenite.

Distinguishing Features—Colorless, isotropic, and equant crystals.

Similar Minerals—Resembles leucite, sodalite, and zeolites, but leucite is often twinned, sodalite has different cleavage, and zeolites show low-order interference colors.

Properties and Interference Figure—Cubic; isotropic; $n = 1.486$.

Color—Colorless.

Form—Usually equant crystals appearing octagonal or rounded; less commonly as irregular masses in ground-mass.

Cleavage—Imperfect cubic.

Relief—Moderate.

Interference Colors—Isotropic but may show extremely low interference colors.

Box 11. Key to Crystal Drawings in Part II

(Note that these drawings are only schematic for minerals with variable compositions.)

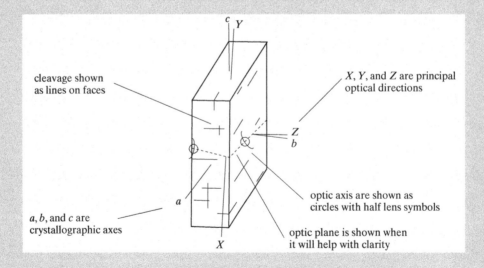

Nepheline $(Na,K)AlSiO_4$

Plates—13c–d, 15c–d

Occurrence—Nepheline is characteristic of some Si-poor igneous rocks such as syenite. It is associated with many minerals including feldspars, apatite, sodalite, zircon, and biotite.

Distinguishing Features—Colorless to cloudy, very low birefringence (gray interference colors), poor cleavage, uniaxial $(-)$.

Similar Minerals—Superficially similar to orthoclase, which is biaxial and has better cleavage. Occasionally confused with scapolite or melilite. Quartz has greater birefringence and is uniaxial $(+)$. Apatite has greater relief. Topaz is biaxial.

Properties and Interference Figure—Hexagonal; uniaxial $(-)$; $\omega = 1.532$–1.547, $\varepsilon = 1.529$–1.542, $\delta = 0.003$–0.005. Basal sections exhibit a uniaxial $(-)$ figure with no isochromes.

Color—Colorless to cloudy.

Form—Often anhedral, but phenocrysts are short prismatic or blocky crystals with rectangular or hexagonal outlines.

Cleavage—Several fair to poor cleavages; rarely seen in thin sections.

Relief—Very low.

Interference Colors—Up to first-order gray.

Extinction and Orientation—Parallel for rectangular sections; basal sections remain extinct in all orientations.

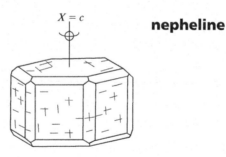

nepheline

Leucite $KAlSi_2O_6$

Plate—15e–f

Occurrence—Leucite is a rare mineral found in Si-poor, K-rich volcanic rocks. It is never found with quartz.

Distinguishing Features—Occurs as phenocrysts in lavas, colorless, extremely low birefringence, complex polysynthetic twinning, enhedral crystals have distinctive 8-sided shape.

Similar Minerals—Sodalite and hercynite are isotropic and have no twin lamellae; sodalite has lower relief; analcime lacks twinning.

Properties and Interference Figure—Tetragonal; uniaxial (+); $\omega = 1.508$, $\varepsilon = 1.509$, $\delta = 0.001$.

Color—Colorless.

Form—Normally euhedral crystals, sometimes with octagonal sections, often containing inclusions giving a concentric or radial pattern.

Cleavage—Poor dodecahedral cleavage is rarely seen in thin sections.

Relief—Low.

Interference Colors—Extremely low birefringence results in very low-order interference colors; sometimes leucite appears isotropic, but anisotropy can be detected by inserting an accessory plate.

Extinction and Orientation—Often characterized by wavy extinction.

Twinning—Complex polysynthetic twins (often showing good birefringence) may occur in three directions, sometimes appearing similar to microcline.

Zeolites (heulandite, stilbite, natrolite, chabazite, and others)

Variable, typically hydrated (Na, Ca, K) aluminosilicates

Plates—stilbite: 28a–b

Occurrence—There are many zeolites; most occur in sedimentary rocks, altered volcanics, and low-grade metamorphic rocks. Larger, more spectacular, samples are found in vugs in basaltic volcanic rocks.

Distinguishing Features—Colorless, first-order interference colors, with low to moderate relief. If coarse, they may show one good or perfect cleavage. Zeolites may form fibrous, radiating, or bladed crystals, but are typically very fine grained unless in vugs.

Similar Minerals—All zeolites are somewhat similar, and they may be difficult or impossible to tell apart. Analcime, sometimes considered a zeolite, may occur as a primary mineral in igneous rocks.

Properties and Interference Figure—Zeolites may be orthorhombic, hexagonal, or monoclinic and so may be biaxial or uniaxial. Optical sign and indices of refraction vary with species and composition.

Color—Colorless.

Form—Zeolites display many forms, including tabular crystals, granular aggregates, sheaf-like or radiating bundles, columns, and fibers. Form can be a key property for distinguishing among the many zeolites.

Cleavage—Often shows one good or perfect cleavage.

Relief—Low to moderate.

Interference Colors—Highest-order colors are first-order white. Cleavage flakes have extremely low birefringence.

Sodalite $Na_3Al_3Si_3O_{12} \cdot NaCl$

Occurrence—Sodalite is commonly associated with nepheline, leucite, or cancrinite in Si-poor, alkali rich, igneous rocks.

Distinguishing Features—Clear to gray, poor cleavage, and isotropic; borders are sometimes dark and show better cleavage than cores.

Similar Minerals—Sodalite resembles analcime, but analcime typically occurs as a secondary mineral and has poor cleavage. Haüyne, another typically isotropic blue mineral, is similar to sodalite but has higher relief and less prominent cleavage. Fluorite has lower relief and octahedral cleavage.

Properties and Interference Figure—Cubic; isotropic; $n = 1.485$.

Color—Clear to gray to blue.

Form—Euhedral crystals appear hexagonal, but much sodalite is anhedral.

Cleavage—Poor, cubic cleavage may show on thin edges of grains.

Relief—Low.

Scapolite   **Solid Solutions of Na$_4$(AlSi$_3$O$_8$)$_3$Cl (marialite)**
and Ca$_4$(Al$_2$Si$_2$O$_8$)$_3$(CO$_3$,SO$_4$) (meionite)

Plate—33a–b

Occurrence—Scapolite is a metamorphic mineral found in marbles, mafic gneisses, or amphibolites.

Distinguishing Features—Colorless, interference colors may be up to second or third order blue depending on composition, parallel extinction, lack of twinning, and uniaxial (−).

Similar Minerals—Similar to plagioclase but not twinned, usually higher birefringence, and feldspars have oblique extinction. Quartz is uniaxial (+). Cordierite is biaxial.

Properties and Interference Figure—Tetragonal; uniaxial (−); ω = 1.539–1.596, ε = 1.537–1.557, δ = 0.002–0.004. Basal sections give a uniaxial (−) figure, longitudinal sections give a flash figure.

Color—Clear; sometimes yellowish.

Form—Columnar, often in aggregates.

Cleavage—Several good to poor, at 90° in a cross section.

Relief—Low to fair.

Interference Colors—Interference colors vary with composition and may be up to second- or third-order blue.

Extinction and Orientation—Normally parallel.

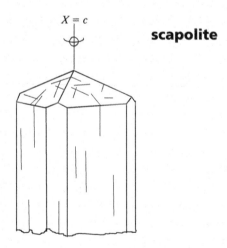

$X = c$

scapolite

Beryl $Be_3Al_2Si_6O_{18}$

Plate—1b

Occurrence—Beryl is found in granitic rocks, most notably pegmatites. It is also found in schists and rare ore deposits.

Distinguishing Features—Colorless; white-gray-yellow interference colors; hexagonal basal sections; often contains fluid inclusions; uniaxial $(-)$.

Similar Minerals—Beryl resembles apatite, which has greater relief, and resembles quartz, which has higher birefringence and is uniaxial $(+)$. Nepheline has lower relief. Topaz is biaxial.

Properties and Interference Figure—Hexagonal; uniaxial $(-)$, very rarely biaxial; $\omega = 1.567$–1.594, $\varepsilon = 1.563$–1.586, $\delta = 0.004$–0.008. Basal sections yield a uniaxial $(-)$ figure with no isochromes.

Color—Normally colorless; occasionally pale yellow or greenish.

Form—Basal sections may appear hexagonal; longitudinal sections are prismatic and may also be massive.

Cleavage—One poor cleavage seen in longitudinal sections.

Relief—Moderate.

Interference Colors—Maximum interference colors are first-order weak yellow.

Extinction and Orientation—Parallel extinction in longitudinal sections; length slow.

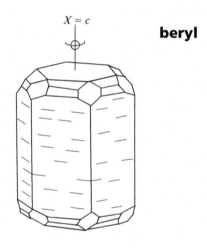

$X = c$

beryl

Cordierite $(Mg,Fe)_2Al_4Si_5O_{18}$

Plates—31c–f, 32a–b

Occurrence—Most cordierite is a product of high-grade metamorphism of aluminous rocks. Rare occurrences in igneous rocks have been reported.

Distinguishing Features—Colorless; first-order colors up to weak yellow; biaxial; pseudohexagonal basal sections; often twinned; sometimes alters to grayish green or yellow fine-grained pinite (sericite). Cordierite may contain pleochroic halos around zircon inclusions, or may contain sillimanite needles.

Similar Minerals—Cordierite may be mistaken for quartz, but quartz is uniaxial (+) and does not alter. Twinned cordierite may look like plagioclase, but cordierite twins are often not perfectly formed.

Properties and Interference Figure—Orthorhombic; biaxial (−) or, more rarely (+), 2V generally 65–90°; $\alpha = 1.530–1.560$, $\beta = 1.535–1.574$, $\gamma = 1.538–1.578$, $\delta = 0.008–0.018$.

Color—Normally colorless; sometimes cloudy; sometimes pleochroic (X = pale orange, Y = light blue-violet, Z = pale blue-violet).

Form—Often anhedral, but may have pseudohexagonal basal sections with prismatic longitudinal sections.

Cleavage—One fair, one poor; usually not obvious in thin sections.

Relief—Low.

Interference Colors—Low birefringence yields first-order white-gray-yellow interference colors.

Extinction and Orientation—Parallel.

Twinning—Cyclic twins give basal sections their pseudohexagonal appearance; longitudinal sections may show lamellae similar to plagioclase.

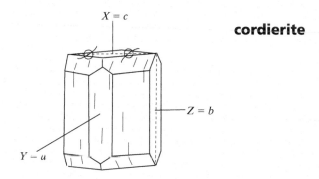

cordierite

▶B. MICAS AND OTHER SHEET SILICATES

Serpentine (chrysotile, lizardite, antigorite) $Mg_6Si_4O_{10}(OH)_8$

Plates—antigorite: 25a–b

Occurrence—Serpentine is most commonly a secondary mineral in mafic and ultramafic igneous rocks, formed by alteration of olivine or pyroxene. It is often associated with fine-grained magnetite or chromite. Serpentine is also found in marbles, associated with carbonates, forsterite, dolomite, or magnesite. It commonly occurs as replacements for individual grains, in massive aggregates, or in veins.

Distinguishing Features—Colorless to green; low birefringence (maximum interference colors are first-order yellow); may be platy or fibrous (chrysotile), normally an alteration product, often as pseudomorphs after mafic minerals; typically appears as an aggregate of small anhedral crystals and may exhibit a coarse "alligator skin" texture.

Similar Minerals—Chlorite is usually biaxial (+), more pleochroic, and displays lower or anomalous interference colors. Micas have greater birefringence. Brucite is uniaxial and displays anomalous interference colors. Chrysotile has lower refractive indices and birefringence than fibrous amphiboles.

Properties and Interference Figure—Monoclinic; biaxial (−) or (+), 2V variable; $\alpha = 1.50–1.55$, $\beta = 1.50–1.60$, $\gamma = 1.50–1.60$, $\delta = 0.006–0.014$. Interference figures are often hard to obtain due to grain size.

Color—Colorless to pale green; may be slightly pleochroic from clear to pale green to yellow green.

Form—Chrysotile forms fibrous veins, mats, or masses. Antigorite is micaceous or flaky, and forms foliated or scaly masses. Lizardite is finer grained and often displays a net-like pattern and undulatory extinction. Serpentine often forms as pseudomorphs after olivine, pyroxene, or other mafic minerals.

Cleavage—One perfect cleavage is visible in antigorite; cleavage is not visible for other serpentines.

Relief—Low.

Interference Colors—Maximum colors are normally first-order gray or white, more rarely first-order yellow. Interference colors may be somewhat anomalous if crystals have a greenish color.

Extinction and Orientation—Extinction is parallel to cleavage; length slow.

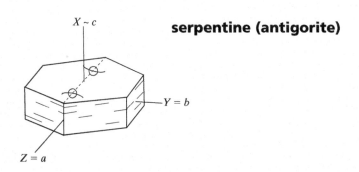

Clay minerals (includes montmorillonite, illite, kaolinite and others)
Hydrous sheet silicates of various compositions

Plate—7e–f

Occurrence—Clays form as end products of weathering and are abundant in a variety of sedimentary rocks and in soils. They also form in secondary veins, and as fine grained alteration products on, or in, coarser grained minerals. In thin sections they may appear as replacement patches in feldspars, micas, and in many other silicates. Incipient alteration to clays can turn normally clear minerals cloudy.

Distinguishing Features—Clay minerals are almost impossible to tell apart in thin sections. They normally appear as massive aggregates of fine scales or shards, and have gray, brown, or earthy colors. Some coarse clays may display up to middle-second-order interference colors, but kaolinite has very low birefringence.

Similar Minerals—All clays are similar, and many other secondary minerals, such a zeolites, may be mistaken for clays. Glauconite, sometimes considered a clay mineral, but more appropriately grouped with micas, is a conspicuous green mineral found in some sandstones (Plate 9e–f).

Properties and Interference Figure—Most clays are monoclinic; biaxial ($-$ or $+$); a few are triclinic or orthorhombic, and a few are hexagonal (uniaxial). Indices of refraction are normally 1.5–1.6, so relief is low. Normally interference figures cannot be obtained because of small grain size.

Color—Normally colorless, but may be gray, brown, tan, or earthy white.

Form—Microcrystalline aggregates of scales or shards.

Relief—Low.

Interference Colors—Up to middle second order.

Pyrophyllite $Al_2Si_4O_{10}(OH)_2$

Occurrence—Pyrophyllite is found in low- and medium-grade metamorphosed shales; more rarely as secondary minerals in felsic volcanics.

Distinguishing Features—Colorless sheet silicate; up to third-order interference colors, normally in patches or aggregates with curved, radial, or distorted crystals; bird's-eye appearance near extinction.

Similar Minerals—Muscovite and talc may be confused with pyrophyllite, but both muscovite and talc have a smaller $2V$, and muscovite crystals are usually not curved or distorted. Pyrophyllite is always associated with quartz, talc is never associated with quartz. Kaolinite has lower birefringence.

Properties and Interference Figure—Triclinic; biaxial $(-)$, $2V = 53°$ to $62°$; $\alpha = 1.554$, $\beta = 1.588$, $\gamma = 1.601$, $\delta = 0.045$–0.048. Cleavage flakes show a centered biaxial figure.

Color—Colorless.

Form—Patches or aggregates with curved, radial, or distorted crystals are common; fine-grained aggregates are also possible.

Cleavage—One perfect cleavage.

Relief—Low to moderate.

Interference Colors—High birefringence yields up to third-order colors in some orientations; sections parallel to cleavage show white or gray colors.

Extinction and Orientation—Parallel (or nearly so) to cleavage and long dimension; slow parallel to good cleavage.

Twinning—May be present but hard to detect.

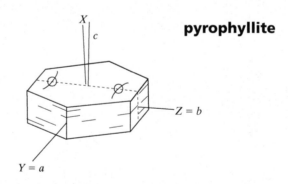

Talc $Mg_3Si_4O_{10}(OH)_2$

Plate—24e–f

Occurrence—Talc is a major mineral in some low-grade metamorphic rocks, including marbles and ultramafic rocks; less commonly, it is a secondary mineral in mafic igneous rocks.

Distinguishing Features—Colorless sheet silicate; third-order interference colors; small *2V*; platy masses or fibrous aggregates with subparallel alignment.

Similar Minerals—Resembles muscovite and pyrophyllite, but both have lower-order interference colors and greater *2V*. Pyrophyllite is always associated with quartz; talc-quartz assemblages are rare. Brucite is uniaxial (+). Gibbsite is biaxial (+), has lower birefringence, and has oblique extinction.

Properties and Interference Figure—Monoclinic; biaxial (−), $2V = 0°$ to $30°$; $\alpha = 1.539$–1.550, $\beta = 1.589$–1.594, $\gamma = 1.589$–1.596, $\delta = 0.046$–0.050.

Color—Colorless.

Form—Bent shreds, thick fibers, or plates are common, often forming fine to coarse subparallel aggregates.

Cleavage—One perfect.

Relief—Low to moderate.

Interference Colors—Maximum interference colors are upper third order; cleavage flakes show white or gray.

Extinction and Orientation—Parallel (or nearly so) to cleavage; length slow.

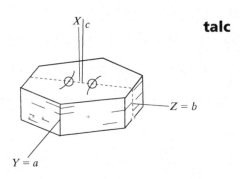

Biotite   **Solid Solutions of KMg$_3$(AlSi$_3$O$_{10}$)(OH)$_2$ (phlogopite)**
                                                                     and KFe$_3$(AlSi$_3$O$_{10}$)(OH)$_2$ (annite)

Plates—1d, 12e–f, 13a–b, 14c–d, 17a–d, 18a–f, 21c–d, 22c–d, 24a–b, 25e–f, 26a–f, 29e–f, 31a–f, 32a–d, 34a–d

Occurrence—Biotite is common in a wide variety of igneous and metamorphic rocks and, to a much lesser extent, in some immature sedimentary rocks.

Distinguishing Features—Brown to yellow, red, or green mica; pleochroic; one excellent cleavage; often shows mottled or bird's eye extinction, or pleochroic halos around small included zircons.

Similar Minerals—Distinguished from hornblende by cleavage and smaller extinction angle; amphiboles have two cleavages at about 60° and 120°. Biotite may occasionally be confused with tourmaline and, less easily, with stilpnomelane. A brown brittle mica, Stilpnomelane, gives a pseudouniaxial figure and has higher relief (Plate 28c–d).

Properties and Interference Figure—Monoclinic (pseudohexagonal); biaxial (−), for most biotite, $2V = 0°$ to 33°; $\alpha = 1.571–1.616$, $\beta = 1.609–1.696$, $\gamma = 1.610–1.697$, $\delta = 0.028–0.081$. Interference figures may appear uniaxial.

Color—Most biotites are pleochroic (X = colorless, light tan, pale greenish brown, or pale green, Y$\cong$Z= brown, olive brown, dark green, or dark red-brown). Phlogopite is colorless to pale brown (Plate 26a–b).

Form—May be in hexagonal plates or tabs; more common as elongate flakes or aggregates, sometimes bent.

Cleavage—One excellent.

Relief—Moderate.

Interference Colors—Strong interference colors range up to second-order red, but they may be hard to see due to color of the mineral; flakes lying on cleavage show very low-order colors.

Extinction and Orientation—Normally shows parallel extinction, but extinction angle may be up to 3°; distorted plates show wavy extinction; cleavage direction is slow.

Twinning—Twins may be present but are usually difficult to see.

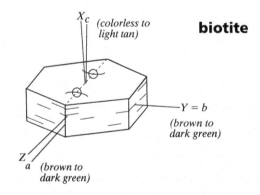

Muscovite $KAl_2(AlSi_3O_{10})(OH)_2$

Plates—5e–h, 9c–d, 18e–f, 25c–d, 26e–f, 29c–d, 30e–f

Occurrence—Muscovite is found in felsic to intermediate igneous rocks, in a wide variety of metamorphic rocks and, less commonly, in immature sedimentary rocks. A fine grained variety of muscovite (sericite) occurs as an alteration product of feldspars and a few other minerals.

Distinguishing Features—Clear mica; second-order interference colors; one excellent cleavage; colorless to pale green color; bird's eye extinction.

Similar Minerals—Some other rarer white micas (paragonite and margarite) and clays have similar properties.

Properties and Interference Figure—Monoclinic; biaxial (−), $2V = 35°$ to $50°$; $\alpha = 1.552–1.770$, $\beta = 1.582–1.619$, $\gamma = 1.588–1.624$, $\delta = 0.036–0.054$. Cleavage fragments yield a nearly centered Bxa (Plate 5e–h).

Color—Usually colorless; may be pale green and slightly pleochroic.

Form—Tabular crystals, flakes, or laths when coarse; scaly aggregates or shreds when fine (sericite: Plates 14a–f, 15c–d, 21a–b, 22c–d, 29a–b, 30a–b).

Cleavage—One excellent cleavage.

Relief—Moderate, may vary slightly with stage rotation.

Interference Colors—Up to second-order yellow or red; sections parallel to cleavage give first-order white.

Extinction and Orientation—Parallel or very small extinction angle; cleavage direction is length slow.

Twinning—Common but hard to detect.

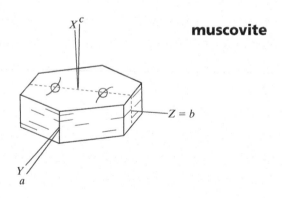

Lepidolite　　　　　　　　　　　　　　　　　　　$K(Li,Al)_{2-3}(AlSi_3O_{10})(OH)_2$

Occurrence—Normally found only in Li-rich pegmatites or granites.

Distinguishing Features—White mica; one good cleavage; up to third-order interference colors; bird's eye extinction.

Similar Minerals—Hard to tell from muscovite, but muscovite typically has a smaller extinction angle, higher relief, and greater birefringence.

Properties and Interference Figure—Monoclinic; biaxial $(-)$, $2V = 0$ to $60°$; $\alpha = 1.524$–1.537, $\beta = 1.543$–1.563, $\gamma = 1.545$–1.566, $\delta = 0.02$–0.04.

Color—Normally colorless; less commonly X = colorless, Y = Z = pink or pale violet.

Form—Tabular or short prismatic crystals; sometimes pseudohexagonal.

Cleavage—One excellent cleavage.

Relief—Low to moderate.

Interference Colors—Strong birefringence yields up to middle third-order colors; sections parallel to cleavage show low-order colors.

Extinction and Orientation—Extinction angle is $0°$ to $6°$; cleavage direction is slow.

Twinning—Common but hard to see.

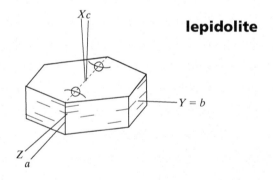

Stilpnomelane $\qquad$ $K_{0.6}(Fe,Mg)_6(AlSi_8)(O,OH)_{27} \cdot 2\text{--}4\ H_2O$

Plate—28c–d

Occurrence—Common in low-grade regional metamorphic rocks. Often associated with chlorite, muscovite, garnet, actinolite, calcite, or epidote.

Distinguishing Features—Dark brown, pale yellow, or green mica-like mineral with strong pleochroism, often showing two cleavages.

Similar Minerals—Hard to tell from biotite, but stilpnomelane's basal cleavage is less perfect than biotite's and is intersected by an additional cleavage; bird's eye extinction is absent or weakly present in stilpnomelane, and stilpnomelane's birefringence may be greater than that of biotite.

Properties and Interference Figure—Monoclinic; biaxial $(-)$, $2V \cong 0$; indices of refraction vary significantly with composition, but typically $\alpha = 1.543\text{--}1.634$, $\beta = 1.576\text{--}1.745$, $\gamma = 1.576\text{--}1.745$, $\delta = 0.030\text{--}0.110$. The biaxial interference figure appears uniaxial because of the small $2V$.

Color—Pale yellow to dark brown or green; pleochroic.

Form—Mica-like habit; may form sheafs or radiating aggregates.

Cleavage—One good and one fair at 90° in properly oriented crystals.

Relief—High, but may vary with stage rotation.

Interference Colors—Birefringence varies with composition; maximum interference colors may be up to sixth order, but the strong mineral color tends to mask them.

Extinction and Orientation—Extinction angle with the principle cleavage is close to 0°; principle cleavage direction is slow.

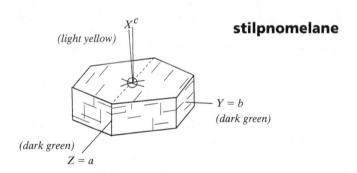

Chlorite   **Variable, typically combinations of Mg$_3$Si$_4$O$_{10}$(OH)$_2$ and Mg(OH)$_2$**

Plates—9c–d, 17a–b, 19a–b, 22e–f, 23a–b, 24c–d, 26c–f

Occurrence—Chlorite is common in low- and medium-grade metamorphic rocks and is a common secondary mineral forming after biotite and other Mg and Fe silicates.

Distinguishing Features—Flake-like crystals or replacement patches; anomalous interference colors (blue, purple, or brown); variable but normally greenish color.

Similar Minerals—There are many varieties of chlorite; the properties listed below are averages.

Properties and Interference Figure—Monoclinic; biaxial (+) or (−), $2V = 0°$ to $50°$; $\alpha = 1.571–1.588$, $\beta = 1.571–1.588$, $\gamma = 1.576–1.597$, $\delta = 0.006–0.020$.

Color—Many varieties, normally colorless to medium green or olive green, more rarely pale brown, may be pleochroic.

Form—Flaky or scaly crystals, or thin to thick tabs; sometimes pseudohexagonal.

Cleavage—One perfect cleavage.

Relief—Moderate to high.

Interference Colors—Very weak, anomalous colors (blue, purple, or brown) are typical; rarely colors up to first-order light yellow.

Extinction and Orientation—Extinction angle varies from $0°$ to about $10°$; crystals showing cleavage are length fast.

Twinning—Polysynthetic twinning is common but generally hard to see.

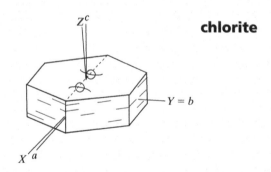

chlorite

Prehnite $Ca_2Al(AlSi_3O_{10})(OH)_2$

Occurrence—Prehnite may be a low-grade metamorphic mineral in mafic rocks but is more commonly an alteration product in basalts and other igneous rocks. Commonly occurs in veins or vugs.

Distinguishing Features—Colorless; second-order colors; sheaf-like aggregates sometimes showing a "bow tie" structure.

Similar Minerals—May be confused with lawsonite, but lawsonite has lower birefringence and higher relief.

Properties and Interference Figure—Orthorhombic; biaxial (+); $2V = 65°$ to $69°$; $\alpha = 1.611$–1.630, $\beta = 1.617$–1.641, $\gamma = 1.632$–1.669, $\delta = 0.021$–0.039.

Color—Colorless.

Form—Sheaf-like aggregates or round globs are typical; sometimes shows a "bow tie" structure when sheaves are pinched in the middle.

Cleavage—One good cleavage.

Relief—Moderate to high.

Interference Colors—Maximum interference colors are low to upper second-order.

Extinction and Orientation—Parallel; may be wavy.

Twinning—Fine polysynthetic twins in two directions may sometimes be seen.

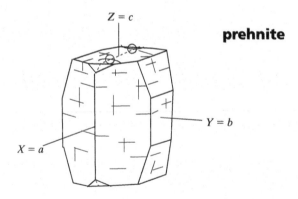

prehnite

▶C. PYROXENES AND PYROXENOIDS

Orthopyroxene

Mg₂Si₂O₆ (enstatite)
(Mg,Fe)₂Si₂O₆ (hypersthene)
Fe₂Si₂O₆ (ferrosilite)

$Mg_2Si_2O_6$ (enstatite)
$(Mg,Fe)_2Si_2O_6$ (hypersthene)
$Fe_2Si_2O_6$ (ferrosilite)

Plates—21c–d, 32a–d

Occurrence—Common in mafic and ultramafic igneous rocks and their high-grade metamorphosed equivalents.

Distinguishing Features—High relief; light green or pleochroic (light green-light pink) color; low birefringence; near 90° cleavage; parallel extinction; sometimes augite exsolution lamellae give a schiller structure. Orthopyroxene has somewhat variable properties depending on composition, and may show compositional zoning.

Similar Minerals—Clinopyroxenes have greater birefringence and inclined extinction. Zoisite is usually length fast and has lower birefringence. Kyanite has oblique extinction. Andalusite is length fast and has lower relief. Amphibole has different cleavage.

Properties and Interference Figure—Orthorhombic; biaxial (+) or (−); $2V = 53°$ to $90°$; $\alpha = 1.650$–1.715, $\beta = 1.653$–1.728, $\gamma = 1.658$–1.731, $\delta = 0.008$–0.022. See graph below. Basal sections reveal a biaxial figure with a moderate to large $2V$.

Color—Colorless or pale pink to green; hypersthene is pleochroic (X = reddish brown or pink, Y = yellow-green, Z = light to gray green).

Form—Well developed crystals are prismatic with square or "stop sign" shaped basal sections.

Cleavage—Basal sections show two good cleavages at nearly right angles; longitudinal sections show one good cleavage.

Relief—Moderately high to high.

Interference Colors—Maximum interference color is mid to upper first-order; more rarely up to second-order blue.

Extinction and Orientation—Extinction is parallel in most sections; cleavage traces are length slow.

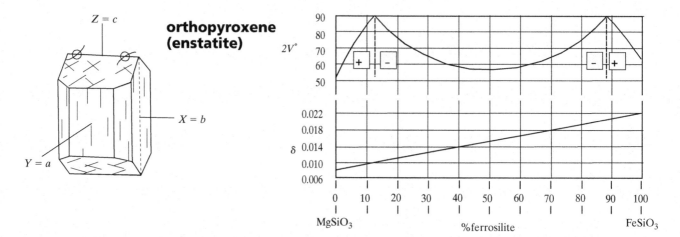

The $2V$ and birefringence (δ) of Mg-Fe orthopyroxene can be used to estimate composition. Note that very Mg-rich and very Fe-rich orthopyroxenes are are biaxial (+); all intermediate compositions are biaxial (−).

Diopside

CaMgSi₂O₆

Plates—20e–f, 21e–f, 26a–b

Occurrence—Found in mafic and ultramafic igneous rocks and their metamorphosed equivalents; also found in some marbles and marls.

Distinguishing Features—High relief, green color, upper second-order interference colors, near 90° cleavage, and a large extinction angle.

Similar Minerals—Orthopyroxene has lower birefringence and parallel extinction. Augite is more deeply colored.

Properties and Interference Figure—Monoclinic; biaxial (+); $2V = 56°$ to $63°$; $\alpha = 1.665$, $\beta = 1.672$, $\gamma = 1.696$, $\delta = 0.031$.

Color—Colorless or pleochroic green of various shades.

Form—Short prismatic subhedral crystals are common; well-formed cross sections may be four- or eight-sided.

Cleavage—Two good cleavages at 87° and 93° can be seen in basal section; one cleavage is seen in longitudinal section.

Relief—Fairly high.

Interference Colors—Maximum interference colors are upper second-order.

Extinction and Orientation—Extinction angle varies up to 44° in longitudinal section depending on orientation and composition; symmetrical extinction in basal sections.

Twinning—Simple and polysynthetic twins are common.

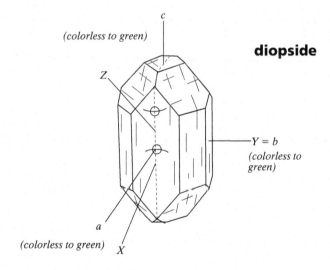

Augite   **(Ca,Fe,Mg,Na)(Mg,Fe,Al)(Si,Al)$_2$O$_6$**

Plates—4a–b, 13e–f, 19e–f, 20a–d, 22a–b

Occurrence—The most common pyroxene found in mafic and intermediate rocks and their metamorphic equivalents.

Distinguishing Features—High relief, normally colorless, pale green or purplish brown, middle second-order interference colors, near 90° cleavage, and a large extinction angle. Augite may contain exsolution lamellae of orthopyroxene, and may show compositional zoning.

Similar Minerals—Sometimes augite is difficult to tell from diopside, but diopside has slightly greater birefringence and is less deeply colored. Orthopyroxene has parallel extinction and lower birefringence. Aegirine and aegirine-augite are related clinopyroxenes characterized by strong green, yellow, or brown coloration.

Properties and Interference Figure—Monoclinic; biaxial (+); $2V = 25°$ to $60°$; $\alpha = 1.670$–1.743, $\beta = 1.676$–1.750, $\gamma = 1.694$–1.772, $\delta = 0.024$–0.029. Cleavage flakes often give good optic axis figure.

Color—Colorless, pleochroic in pale greens, or purplish brown.

Form—Often appears as short prismatic crystals with four- or eight-sided cross sections (Plate 20a–d).

Cleavage—Basal sections show two good cleavages at 87° and 93°; longitudinal sections show one good cleavage.

Relief—High.

Interference Colors—Maximum interference colors are middle second-order.

Extinction and Orientation—Maximum extinction angle in longitudinal sections is between 36° and 45° depending on composition; sections that have maximum extinction angle also have maximum birefringence; symmetrical extinction in basal sections.

Twinning—Simple or polysynthetic twins are common; a combination of the two produces a "herringbone" appearance.

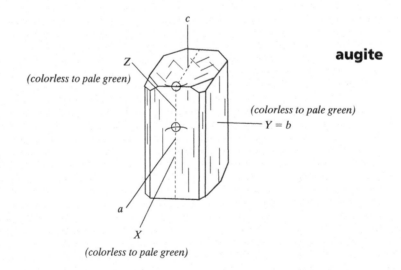

Pigeonite $(Mg,Fe,Ca)_2(Si,Al)_2O_6$

Plate—21a–b

Occurrence—Most often found as a matrix mineral with augite or orthopyroxene in andesite, dacite, basalt, or in diabase.

Distinguishing Features—Typical pyroxene shape and cleavage, moderate birefringence, and small *2V*.

Similar Minerals—Sometimes pigeonite is difficult to distinguish from augite, but augite has a larger *2V* and may occur as phenocrysts and in plutonic rocks. Olivine has higher birefringence and lacks cleavage. Orthopyroxene has lower birefringence and larger *2V*.

Properties and Interference Figure—Monoclinic; biaxial (+); *2V* = 0° to 32°; α = 1.682–1.722, β = 1.684–1.722, γ = 1.705–1.751, δ = 0.023–0.029. Cleavage flakes give off-center figures.

Color—Colorless or pale green or brown, may be pleochroic in pale colors.

Form—Stubby euhedral to anhedral prismatic crystals are typical.

Cleavage—Basal sections show two good cleavages at 87° and 93°; longitudinal sections show one good cleavage.

Relief—High.

Interference Colors—Maximum interference colors are normally first-order red or yellow but may range up to lower second-order.

Extinction and Orientation—Symmetrical extinction in basal sections showing both cleavages; in longitudinal sections, extinction may be parallel or inclined depending on orientation. The maximum extinction angle is about 45° depending on composition; sections that have a maximum extinction angle also have maximum birefringence.

Twinning—Simple or polysynthetic twins are common (Plate 21a–b); combination of the two produces a "herringbone" structure. Exsolution of augite is common and may be confused with twinning.

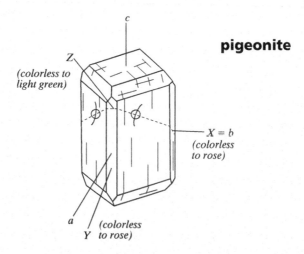

pigeonite

Jadeite   **NaAlSi₂O₆**

Plate—20g–h

Occurrence—A high pressure pyroxene found in metamorphic rocks of the blueschist facies.

Distinguishing Features—Colorless to light green; sometimes pleochroic; maximum interference colors may be first order, but anomalous blue interference colors are common; two cleavages at nearly 90°. Composition and properties vary somewhat.

Similar Minerals—Columnar habit, extinction angle and (sometimes) anomalous interference colors help distinguish jadeite from diopside and augite. Actinolite has a smaller extinction angle.

Properties and Interference Figure—Monoclinic; biaxial (+); $2V$ = 68° to 72°; α = 1.654, β = 1.657, γ = 1.665, δ = 0.011–0.016.

Color—Usually colorless, sometimes pleochroic (X = pale green, Y = colorless, Z = yellowish or yellow-green).

Form—Rarely euhedral; typically granular or columnar or patchy aggregates.

Cleavage—Two good cleavages at 87° and 93° are seen in some sections; one good cleavage in others.

Relief—High.

Interference Colors—Maximum interference colors may be upper first order or lower second order, but anomalous blue interference colors are common.

Extinction and Orientation—Maximum extinction angle in longitudinal sections is 30° to 45°, depending on composition.

Twinning—Simple twins may be present.

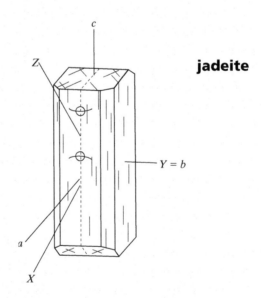

jadeite

Wollastonite   CaSiO$_3$

Plate—21e–f

Occurrence—Found in high-grade siliceous marbles and related calcareous metamorphic rocks.

Distinguishing Features—Colorless; first-order interference colors; near parallel extinction; tends to fracture during thin sectioning.

Similar Minerals—Tremolite resembles wollastonite but has oblique extinction and amphibole cleavage. Wollastonite may resemble sillimanite, but sillimanite is biaxial (−) and does not occur in marbles.

Properties and Interference Figure—Triclinic; biaxial (−); $2V = 36°$ to $42°$; $\alpha = 1.620$, $\beta = 1.632$, $\gamma = 1.634$, $\delta = 0.014$.

Color—Colorless.

Form—Columnar or fibrous aggregates are typical; cross sections may be nearly square.

Cleavage—Two perfect, one good.

Relief—Fairly high.

Interference Colors—Maximum colors are about first-order orange.

Extinction and Orientation—Parallel or near parallel extinction in longitudinal section; may be length slow or length fast depending on orientation.

Twinning—Multiple twins are common.

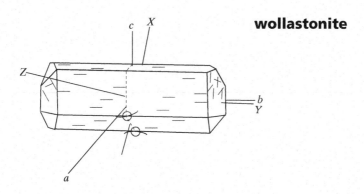

wollastonite

▶D. AMPHIBOLES

Anthophyllite $(Mg,Fe)_7Si_8O_{22}(OH)_2$

Occurrence—Found in low- to medium-grade Mg-rich metamorphic rocks, especially in serpentinites.

Distinguishing Features—Colorless or nearly colorless; up to low second-order colors; amphibole cleavage and habit. Often occurs as long prisms or fibers, extinction parallel to long direction.

Similar Minerals—Resembles tremolite-actinolite and other clinoamphiboles, but they have oblique extinction. Gedrite is similar to anthophyllite but is biaxial (−) and typically pleochroic gray or green. Zoisite has greater relief and lower birefringence.

Properties and Interference Figure—Orthorhombic; biaxial (+) or (−); $2V = 65°$ to $90°$; $\alpha = 1.598–1.647$, $\beta = 1.616–1.651$, $\gamma = 1.623–1.664$, $\delta = 0.017–0.026$.

Color—Colorless or pale brown or green; may be slightly pleochroic.

Form—Long prismatic crystals have diamond-shaped cross sections displaying amphibole cleavage.

Cleavage—Two excellent cleavages at about 55° and 125° can be seen in cross sections; one cleavage in longitudinal section.

Relief—Moderate to high.

Interference Colors—Up to low second-order.

Extinction and Orientation—Parallel extinction in longitudinal sections; symmetrical extinction in cross sections.

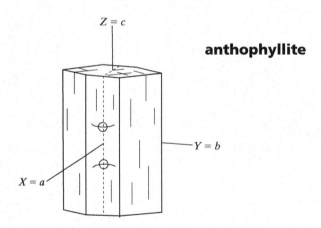

Cummingtonite-grunerite

Solid solution between Mg$_7$Si$_8$O$_{22}$(OH)$_2$ (Mg-cummingtonite) and Fe$_7$Si$_8$O$_{22}$(OH)$_2$ (grunerite)

Plates

cummingtonite: 24a–b
grunerite: 24c–d

Occurrence—Mg-rich amphiboles are found in medium-grade mafic and marly metamorphic rocks, especially amphibolites. Grunerite is found in metamorphosed Fe-rich sediments, including banded iron formations.

Distinguishing Features—Colorless or weakly colored; second-order to lower third-order interference colors; typically parallel or radiating aggregates of long prismatic crystals or fibers; amphibole cleavage, polysynthetic twinning; sometimes contains exsolution lamellae of hornblende.

Similar Minerals—Distinguished from anthophyllite by inclined extinction. Actinolite is biaxial (−), typically more deeply colored, and has lower relief. Chrysotile has lower relief and birefringence. Cummingtonite may be difficult to distinguish from tremolite.

Properties and Interference Figure—Monoclinic; cummingtonite is biaxial (+), grunerite is biaxial (−); $2V = 80°$ to $90°$; $\alpha = 1.644$–1.685, $\beta = 1.657$–1.709, $\gamma = 1.614$–1.728, $\delta = 0.030$–0.043.

Color—Mg-rich varieties are colorless; Fe-rich varieties may be pleochroic in green and yellow.

Form—Commonly parallel to radiating prismatic or fibrous crystals with diamond-shaped cross sections.

Cleavage—Two good cleavages at 56° and 124° may be seen in a cross section; one cleavage is seen in a longitudinal section.

Relief—Moderate to high.

Interference Colors—Maximum colors are middle second order to lower third order.

Extinction and Orientation—Maximum extinction angle varies from about 10° to 20° in longitudinal sections, depending on composition.

Twinning—Polysynthetic twinning is common.

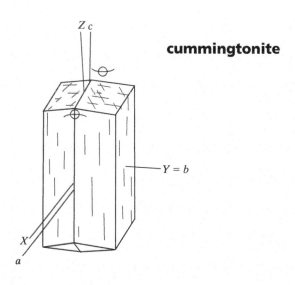

cummingtonite

Tremolite-actinolite-ferroactinolite $Ca_2Mg_5Si_8O_{22}(OH)_2$ (tremolite)
$Ca_2(Mg,Fe)_5Si_8O_{22}(OH)_2$ (actinolite)
$Ca_2(Fe,Mg)_5Si_8O_{22}(OH)_2$ (ferroactinolite)

Plates

tremolite: 23c–f, 24e–f, 33e–f
actinolite: 23a–b

Occurrence—Tremolite is found in marbles; actinolite is found in metamorphosed intermediate to mafic rocks.

Distinguishing Features—Colorless (tremolite) to pale green (actinolite); perhaps faintly pleochroic with typical amphibole cross section (diamond-shaped) and prismatic or fibrous habit; inclined extinction.

Similar Minerals—Similar to some pyroxenes and pyroxenoids, but cleavage is diagnostic. Similar to some orthoamphiboles, but they have parallel extinction in longitudinal section. Wollastonite has a smaller $2V$. Hornblende is more deeply colored. Tremolite is occasionally confused with other light-colored amphiboles.

Properties and Interference Figure—Monoclinic; biaxial $(-)$; $2V = 74°$ to $85°$; $\alpha = 1.608$–1.688, $\beta = 1.618$–1.699, $\gamma = 1.630$–1.704, $\delta = 0.016$–0.030. Longitudinal sections with low-order interference colors give best figures.

Color—Tremolite is colorless, actinolite is colorless to pale green and pleochroic (X = pale yellow-green, Y = light green, Z = light bluish green).

Form—Long prismatic crystals with diamond-shaped cross sections, either individually, in non-aligned masses, or in columnar aggregates; some varieties are asbestiform.

Cleavage—Two good cleavages at $56°$ and $124°$ can be seen in a cross section; one cleavage is seen in a longitudinal section.

Relief—Moderate to high.

Interference Colors—Maximum colors are low to middle second order.

Extinction and Orientation—Maximum extinction angle in longitudinal sections is $10°$ to $20°$; symmetrical extinction in cross sections; longitudinal sections are length slow.

Twinning—Simple twins are common, polysynthetic twins less so.

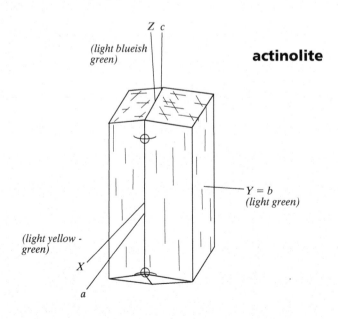

Hornblende $(Na,K)_{0-1}(Ca,Na,Fe,Mg)_2(Mg,Fe,Al)_5(Si,Al)_8O_{22}(OH)_2$

Plates—12c–d, 14c–d, 17c–f, 22a–f, 28e–f, 34a–b

Occurrence—Found in a wide variety of igneous and metamorphic rocks.

Distinguishing Features—Deep color and pleochroism; characteristic amphibole cleavage and habit. Properties are variable due to variable composition. Rarely, hornblende contains exsolution lamellae. Crystals generally short prisms, more rarely elongate.

Similar Minerals—Can be distinguished from augite by cleavage, pleochroism, and extinction angle; may resemble biotite, but biotite has only one cleavage and near parallel (sometimes bird's eye) extinction. Allanite, a rare earth bearing mineral similar to epidote, has parallel extinction and only one cleavage. The Na-amphiboles glaucophane, riebeckite, and crocidolite (which is asbestiform) have characteristic pleochroic blue or violet colors.

Properties and Interference Figure—Monoclinic; biaxial $(-)$; $2V = 52°$ to $85°$; $\alpha = 1.614$–1.675, $\beta = 1.618$–1.691, $\gamma = 1.633$–1.701, $\delta = 0.019$–0.026. See drawing of actinolite or previous page 74.

Color—Pleochroic in various shades of green and brown and, less commonly, yellow or red-brown. Concentric or patchy color zoning may be present.

Form—Typical amphibole habit: prismatic crystals with imperfect diamond-shaped cross sections; longitudinal sections may appear as laths in thin sections.

Cleavage—Two good cleavages at $56°$ and $124°$ (Plate 22c–d).

Relief—Moderate to high.

Interference Colors—Maximum colors are about middle second order, but they may be masked by the deep color of the mineral.

Extinction and Orientation—In longitudinal sections, the maximum extinction angle varies from about $12°$ to $30°$ depending on composition; in cross sections, extinction is symmetrical to the two cleavages.

Twinning—Simple twins are relatively common (Plate 22c–d).

Na Amphiboles   Na$_2$Mg$_3$Al$_2$(Si$_8$O$_{22}$)(OH)$_2$ (Glaucophane)
Na$_2$(Mg,Fe)$_3$(Al,Fe)$_2$(Si$_8$O$_{22}$)(OH)$_2$ (Crossite)
Na$_2$Fe$_3$Fe$_2$(Si$_8$O$_{22}$)(OH)$_2$ (Riebeckite)

Plates—20g–h, 22e–f, 27e–f

Occurrence—These blue amphiboles are found in high-pressure schists (blueschists) and gneisses.

Distinguishing Features—Distinguished by presence in high pressure rocks, amphibole cleavage, a distinctive blue color, and pleochroism. The blue amphiboles may be difficult to tell apart. Glaucophane and riebeckite form a solid solutions series, and crossite refers to compositions between the two, so properties grade into each other. Riebeckite is usually darker blue than glaucophane, may be biaxial (+), and normally has a *2V* > 45°. Occasionally blue amphibole is confused with blue tourmaline, which is uniaxial.

Properties and Interference Figure—Monoclinic; glaucophane is biaxial (−), *2V* = 10° to 45°; riebeckite is biaxial (+ or −), *2V* normally 45° to 90°; α = 1.606–1.701, β = 1.622–1.711, γ = 1.627–1.717, δ = 0.006–0.029.

Color—Pleochroic from colorless to deep blue or violet. Glaucophane: X = colorless or pale blue; Y = lavender-blue, bluish green; Z = blue, greenish blue, or violet. Riebeckite: X = dark blue; Y = indigo blue or gray-blue; Z = blue, yellow-green, or yellow-brown.

Form—Long slender crystals with diamond-shaped cross sections are typical; less commonly fibrous, or in columnar aggregates, or in foliated masses.

Cleavage—Typical amphibole cleavage is visible in a cross section.

Relief—Moderate to moderately high.

Interference Colors—Maximum colors are middle to upper first order but may be anomalous due to the color of the mineral.

Extinction and Orientation—Extinction angle to cleavage is up to 6°; glaucophane is length slow; riebeckite is length fast.

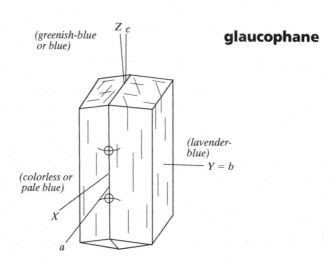

▶E. RING SILICATES

Tourmaline (Na,Ca)(Li,Fe,Mg,Al)Al₆(BO₃)₃Si₆O₁₈(OH)₄

Plate—18a–b

Occurrence—A common accessory mineral in granitic igneous rocks and in some metamorphic rocks, especially mica schists; also found as a major mineral in some pegmatite and (rarely) in immature sedimentary rocks.

Distinguishing Features—Deep color often masks interference colors, moderately high birefringence when visible, and hexagonal cross sections. No cleavage.

Similar Minerals—Tourmaline may occasionally be confused with biotite or hornblende, but lacks cleavage, is more deeply colored, and has a triagonal or hexagonal prismatic habit. Some Na-rich pyroxenes look superficially like tourmaline but have greater relief and birefringence. Dumortierite, another often strongly colored boron mineral, resembles some varieties of tourmaline.

Properties and Interference Figure—Hexagonal (rhombohedral); uniaxial (−); ω = 1.639–1.692, ε = 1.620–1.657, δ = 0.019–0.035. Basal sections yield a uniaxial (−) figure showing one to several isochromes.

Color—Variable: may be black, brown, green, blue, yellow, red, or pink; usually strongly pleochroic and may show color zonation.

Form—Stubby columnar or prismatic crystals with rounded triangular or hexagonal cross sections are typical; columnar or fibrous radiating aggregates are common.

Cleavage—Absent.

Relief—High.

Interference Colors—Cross sections show no birefringence; other sections may exhibit up to upper second-order colors, but interference colors are often masked by color of the mineral.

Extinction and Orientation—Extinction is parallel in most sections; most tourmaline crystals are length fast.

▶F. GARNET, OLIVINE, AND OTHER ISOLATED TETRAHEDRAL SILICATES

Garnet $(Fe,Mg,Ca,Mn)_3(Al,Fe)_2Si_3O_{12}$

Plates—1a, 1d, 21e–f, 24a–b, 28c–f, 30e–f, 31e–f

Occurrence—Garnet is found in a wide variety of metamorphic rocks and in some igneous rocks. Mg-garnet (pyrope) is typically found in ultramafic rocks; Fe-garnet (almandine) is common in mica schists and gneisses; Mn-garnet (spessartine) is found in granites and pegmatites; Ca-garnet (grossular or andradite) is found in skarns and marbles.

Distinguishing Features—High relief, isotropic, colorless but often having a pale tinge, irregular fracture, and often with inclusions.

Similar Minerals—Spinel, also isotropic, may occasionally be confused with garnet. Perovskite $(CaTiO_3)$ resembles garnet but has much higher relief.

Properties and Interference Figure—Cubic; isotropic; $n = 1.71—1.87$.

Color—Colorless or pale pink, red, brown, green, or gray; rarely darker colors; sometimes zoned. Pyralspite garnets (solid solutions of pyrope-almandine-spessartine) are normally red, but the color may not show in thin section.

Form—Euhedral crystals are six- or eight-sided; irregular polygons or subhedral to anhedral crystals are also common.

Cleavage—None, but irregular fractures are common.

Relief—Very high.

Interference Colors—Although garnet is isotropic, some rare varieties show weak retardation.

Olivine   Solid solution of Mg$_2$SiO$_4$ (forsterite) and Fe$_2$SiO$_4$ (fayalite)

Plates—6a–c, 15a–b, 19a–f

Occurrence—Olivine is a primary mineral in many igneous rocks, especially mafic ones. It is also found in metamorphic rocks including marble and gneiss.

Distinguishing Features—Colorless, upper second-order interference colors, high relief, irregular fracturing, lack of cleavage, alteration to iddingsite (distinctive red to yellow-brown; see, for example, Plate 19c), and serpentine or chlorite.

Similar Minerals—Diopside is occasionally confused with olivine, but diopside has better cleavage, and lower relief and birefringence. Epidote has a different color and often displays anomalous interference colors. Chondrodite is similar in some respects but is usually pleochroic from neutral to brown or red-brown.

Properties and Interference Figure—Orthorhombic; biaxial (+) or (−); *2V* = 47–54° (Fe-rich) to 85–90° (Mg-rich); α = 1.641–1.835, β = 1.651–1.877, γ = 1.670–1.886, δ = 0.035–0.051.

Color—Colorless, or (Fe-rich) very faintly colored pleochroic yellow.

Form—Normally anhedral but phenocrysts may have a six-sided polygonal outline.

Cleavage—One good, one poor, rarely seen in thin sections.

Relief—Fairly high.

Interference Colors—Maximum colors are upper second order.

Extinction and Orientation—Parallel to cleavage and crystal outlines.

Twinning—Rare and usually poorly developed.

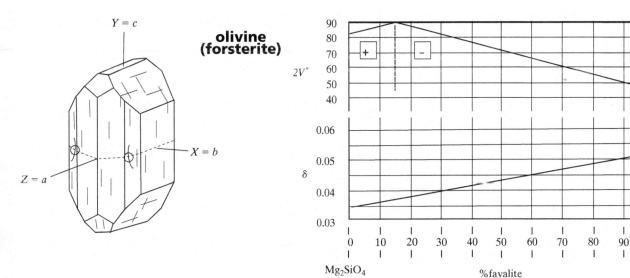

The *2V* and birefringence (δ) of olivine can be used to estimate composition. Note that very Mg-rich olivines are biaxial (+); all others are biaxial (−).

Kyanite Al_2SiO_5

Plate—30a–b

Occurrence—Primarily a metamorphic mineral found in medium or high pressure mica schists and gneisses.

Distinguishing Features—Colorless to pale blue; high relief; normally shows upper first-order interference colors; up to 30° extinction angle depending on orientation; biaxial (−); well developed cleavage.

Similar Minerals—Similar to its polymorphs andalusite and sillimanite but can be distinguished by a large extinction angle. Orthopyroxene has a smaller $2V$, clinozoisite has different cleavage, and lawsonite is biaxial (+) and has different cleavage.

Properties and Interference Figure—Triclinic; biaxial (−); $2V = 82°$; $\alpha = 1.711$, $= 1.721$, $\gamma = 1.728$, $\delta = 0.017$.

Color—Typically colorless to pale grayish blue; occasionally X = colorless, Y = pale violet-blue, Z = pale cobalt blue.

Form—Broad elongate tabs or narrow prisms; often bent.

Cleavage—One perfect cleavage and one good cleavage parallel to long dimension; parting at nearly 85° to the length of a crystal may appear as another cleavage.

Relief—High.

Interference Colors—Up to first-order red.

Extinction and Orientation—Extinction angle in long sections is 0° to 30°; in cross sections, extinction is nearly parallel.

Twinning—Simple, or more rarely polysynthetic, twins are common.

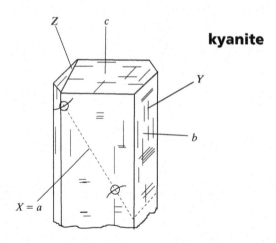

Andalusite Al_2SiO_5

Plate—29a–d

Occurrence—A metamorphic mineral that occurs in micaceous rocks.

Distinguishing Features—Colorless or (rarely) reddish, high relief, low birefringence, and parallel extinction.

Similar Minerals—Distinguished from sillimanite by being length fast, having weaker birefringence, and having a large *2V*. Kyanite has oblique extinction. Colored varieties of andalusite may resemble orthopyroxene, but orthopyroxene is length slow.

Properties and Interference Figure—Orthorhombic; biaxial (−); *2V* = 83° to 85°; α = 1.629–1.640, β = 1.633–1.644, γ = 1.638–1.651, δ = 0.009–0.011.

Color—Colorless, rarely pleochroic pink, rose red or light green.

Form—Euhedral crystals with square outlines or coarse columnar aggregates are typical; inclusions of dark organic matter are common and may form symmetrical cross-like patterns termed chiastolite (Plate 29a–b).

Cleavage—Two cleavages at about 90° in cross sections; one cleavage visible in long sections.

Relief—Fairly high.

Interference Colors—Low birefringence results in maximum colors being first-order yellow.

Extinction and Orientation—Parallel in most sections; symmetrical to cleavage in cross sections; elongate crystals are length fast.

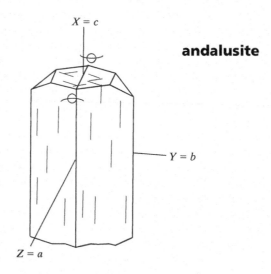

andalusite

Sillimanite Al_2SiO_5

Plates—29c–f, 31c–f

Occurrence—A metamorphic mineral found in high-grade aluminous schists and gneisses.

Distinguishing Features—Colorless; high relief; up to second-order blue colors; needle-like, prismatic, or bladed form with parallel extinction; characteristic square cross sections with one diagonal cleavage; occasionally in a fine fibrous mass (fibrolite).

Similar Minerals—Similar to andalusite, but sillimanite is length slow and has greater birefringence and smaller *2V* than andalusite. Kyanite has greater relief and greater *2V*. Zoisite has lower birefringence but greater relief.

Properties and Interference Figure—Orthorhombic; biaxial (+); $2V = 21°$ to $30°$; $\alpha = 1.659$, $\beta = 1.660$, $\gamma = 1.679$, $\delta = 0.021$. Cross sections show a Bxa, but good figures are hard to obtain because crystals always have at least one short dimension.

Color—Usually colorless; rarely brownish.

Form—Often very small needles or slender prisms; when coarser, cross sections are diamond-shaped or square with one diagonal cleavage; occasionally in a fine grained mass (fibrolite, Plates 29c–d); needles may be bent.

Cleavage—One good cleavage.

Relief—Fairly high.

Interference Colors—Moderate birefringence results in up to second-order blue colors.

Extinction and Orientation—Extinction is parallel to faces in longitudinal sections, symmetrical to faces in cross sections; length slow.

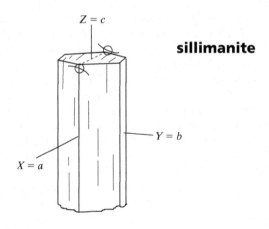

Staurolite $Fe_2Al_9Si_4O_{23}(OH)$

Plate—30c–d

Occurrence—A metamorphic mineral common in medium-grade schists and gneisses.

Distinguishing Features—Pale to strong yellow color, normally pleochroic, usually having many quartz inclusions (Swiss cheese texture, Plate 30c–d); may have six-sided cross sections.

Similar Minerals—Staurolite is sometimes confused with tourmaline (uniaxial). Epidote is lighter colored and more greenish.

Properties and Interference Figure—Monoclinic, biaxial (+); $2V = 79°$ to $90°$; $\alpha = 1.739–1.747$, $\beta = 1.745–1.753$, $\gamma = 1.752–1.762$, $\delta = 0.013–0.015$.

Color—Typically yellow or pleochroic (X = pale yellow, Y = light yellow, Z = reddish yellow).

Form—Euhedral to subhedral crystals are common; short prismatic crystals may have well-formed, six-sided cross sections.

Cleavage—One poor, generally not observed.

Relief—High.

Interference Colors—Maximum colors are first-order yellow to red.

Extinction and Orientation—Normally parallel in long sections; symmetrical in cross sections; length slow.

Twinning—Penetration twins are common but are rarely seen in thin sections.

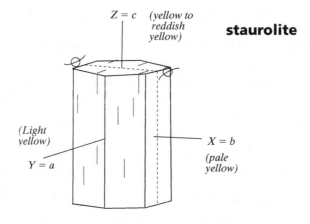

Chloritoid   **$(Fe,Mg)_2Al_4Si_2O_{10}(OH)_4$**

Plate—30e–f

Occurrence—A metamorphic mineral common in low- and medium-grade Fe- and Al-rich schists.

Distinguishing Features—Platy; green or greenish gray color; sometimes with an "hour glass" structure; high relief; first-order interference colors.

Similar Minerals—Resembles chlorite but chloritoid's relief is much higher. Green biotite has a smaller *2V*, higher birefringence, and a single good cleavage. Na-rich amphiboles have amphibole cleavage.

Properties and Interference Figure—Monoclinic; biaxial (+); *2V* = 36° to 63°; α = 1.713–1.728, β = 1.719–1.734, γ = 1.723–1.740, δ = 0.010–0.012.

Color—Very rarely colorless; typically various shades of "military" green, gray and blue; pleochroic (X = gray-green to olive green, Y = gray-blue, Z = colorless, yellowish green or light blue).

Form—Tabular crystals with pseudohexagonal cross sections; inclusions often reveal an "hour glass" pattern.

Cleavage—One perfect and one poor cleavage sometimes give cross sections two cleavages at near to 60°.

Relief—High.

Interference Colors—Maximum colors are middle to upper first order.

Extinction and Orientation—Extinction is 0° to 20°; crystals are length fast.

Twinning—Polysynthetic twins are common.

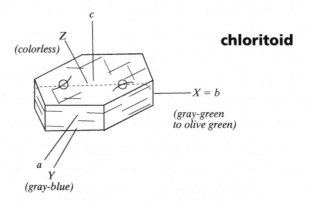

Titanite (sphene) CaTiSiO$_5$

Plates—17c–f, 22e–f, 33a–b

Occurrence—Titanite is an often overlooked accessory mineral in many igneous and metamorphic rocks. It has also been identified in a few sedimentary rocks.

Distinguishing Features—Colorless; very high relief; diamond or distorted six-sided shape with very high-order interference colors that may be difficult to identify.

Similar Minerals—Monazite, a rare phosphate mineral, is similar in some ways to titanite but has lower birefringence. Epidote has lower relief and birefringence and greater *2V*.

Properties and Interference Figure—Monoclinic; biaxial (+); $2V = 23°$ to $50°$; $\alpha = 1.885$–1.921, $\beta = 1.896$–1.927, $\gamma = 1.993$–2.081, $\delta = 0.108$–0.160.

Color—Colorless or very weakly colored (X = almost colorless, Y = pale greenish to yellow, Z = wine yellow).

Form—Euhedral crystals are typical; most often shaped as diamonds, squashed hexagons, or wedges.

Cleavage—One good cleavage, often not seen. Several directions of parting may be mistaken for cleavage.

Relief—Very high.

Interference Colors—Extreme birefringence results in very high-order colors that may appear as white.

Extinction and Orientation—Cross sections display symmetrical extinction but, due to strong dispersion, titanite may show flashes of orange and blue and not go completely extinct.

Twinning—Twins, sometimes polysynthetic, may be present.

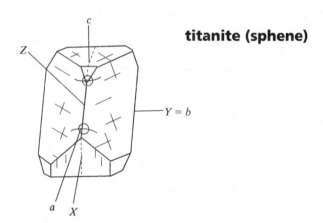

titanite (sphene)

Zircon   ZrSiO$_4$

Plates—18a–d, 32a–b

Occurrence—A widespread accessory mineral in igneous rocks and their metamorphic equivalents. It is also common in sediments and sedimentary rocks.

Distinguishing Features—Small prismatic crystals with high relief and extreme birefringence; often surrounded by pleochroic halos when present as inclusions in other minerals.

Similar Minerals—Sometimes confused with apatite when fine grained, but zircon has much greater birefringence and higher relief. Sometimes mistaken for epidote, monazite, or titanite, which also have high/extreme birefringence.

Properties and Interference Figure—Tetragonal; uniaxial (+); ω = 1.925–1.931, ε = 1.985–1.993, δ = 0.060–0.062. Interference figures show many isochromes.

Color—Colorless, rarely pale brownish or greenish.

Form—Crystals may appear as small prisms with pyramidal terminations or small slender rounded grains.

Cleavage—Absent.

Relief—Very high.

Interference Colors—High birefringence gives up to fourth-order colors.

Extinction and Orientation—Parallel to the long dimension; length slow.

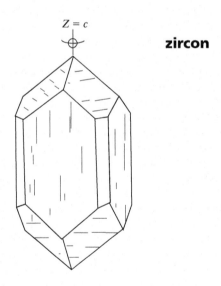

zircon

▶G. PAIRED TETRAHEDRAL SILICATES AND RELATED MINERALS

Lawsonite $CaAl_2Si_2O_7(OH)_2 \cdot H_2O$

Plate—27e–f

Occurrence—A rare mineral found in metamorphic rocks of the blueschist facies.

Distinguishing Features—Clear, pseudorhombic cross sections; high relief; second-order colors; sometimes with polysynthetic twins.

Similar Minerals—Clinozoisite resembles lawsonite but often has anomalous interference colors. Prehnite has greater birefringence; Tremolite has oblique extinction; and Wollastonite has a smaller *2V*.

Properties and Interference Figure—Orthorhombic; biaxial (+); $2V = 79°$ to $85°$; $\alpha = 1.665$, $\beta = 1.674$, $\gamma = 1.685$, $\delta = 0.020$.

Color—Colorless.

Form—Crystals are normally subhedral to euhedral with rhombic cross sections.

Cleavage—Two perfect, two fair; two perpendicular cleavages may show in tabular or elongate section; two "rhombic" cleavages show in cross section.

Relief—Moderately high.

Interference Colors—Maximum are up to second-order blue.

Extinction and Orientation—Symmetrical in cross section; parallel to long dimension.

Twinning—Complex polysynthetic twins are common.

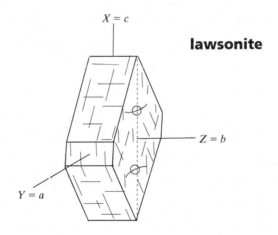

Vesuvianite (Idocrase) $Ca_{10}(Mg,Fe)_2Al_4Si_9O_{34}(OH)_4$

Plate—33c–d

Occurrence—A rare mineral found primarily in contact aureoles associated with impure limestone or dolomite.

Distinguishing Features—Occurrence and association, high relief, low birefringence, and often anomalous interference colors identify vesuvianite.

Similar Minerals—Clinozoisite, epidote, and zoisite are biaxial and usually show cleavage. Epidote has greater birefringence. Andalusite is biaxial. Andalusite and apatite both have lower relief than vesuvianite.

Properties and Interference Figure—Tetragonal; uniaxial (−); $\omega = 1.702\text{–}1.795$, $\varepsilon = 1.700\text{–}1.774$, $\delta = 0.001\text{–}0.021$.

Color—Typically colorless in thin sections; more rarely light green, yellow, or brown. Pleochroism absent or very weak.

Form—Crystals are normally short tetragonal prisms; more rarely, anhedral grains or radial, columnar, or fibrous aggregates.

Cleavage—Several poor cleavages; rarely visible in thin sections.

Relief—High.

Interference Colors—Low or anomalous blue, olive, yellow, or brown.

Extinction and Orientation—Normally, elongate crystals are length fast and display parallel extinction.

Twinning—Absent.

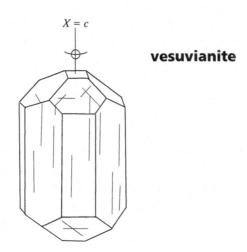

vesuvianite

Epidote and Clinozoisite

$Ca_2(Al,Fe)Al_2Si_3O_{12}(OH)$
$Ca_2Al_3Si_3O_{12}(OH)$

Plates—4c–d, 27a–d

Occurrence—Epidote and clinozoisite are similar in composition and structure. They are common in low- and medium-grade metamorphic rocks, especially amphibolites and marbles. Epidote can also be produced by alteration of feldspars in rocks of many different compositions.

Distinguishing Features—Clinozoisite is normally colorless, but epidote is typically yellowish green; both show high relief and parallel extinction; anomalous blue, green, or yellow interference colors are common, especially for clinozoisite; epidote has greater birefringence than clinozoisite and may show upper second- to third-order colors.

Similar Minerals—Green epidote is distinguished from most clinopyroxenes by cleavage, parallel extinction, and optic sign. Zoisite and clinozoisite have lower birefringence than epidote. Fayalite is normally yellow or yellow-green and shows little cleavage. Vesuvianite is uniaxial (−). Allanite, a rare earth-bearing mineral, is typically pleochroic brown. Piedmontite is an Mn-rich variety of epidote with strong colors (yellow, orange, red, or violet) and marked pleochroism.

Properties and Interference Figure—Monoclinic; epidote is biaxial (−) with $2V = 69°$ to $89°$; clinozoisite is biaxial (+) with $2V = 14°$ to $90°$; indices of refraction are high for both (>1.71 for epidote, >1.67 for clinozoisite), $\delta = 0.015–0.048$ (epidote) or $0.005–0.015$ (clinozoisite). Cleavage fragments often show an optic axis figure.

Color—Clinozoisite is colorless; epidote is normally pleochroic in yellowish green (X = colorless to lemon yellow, Y = greenish yellow, Z = colorless to light yellowish); grains may be color zoned.

Form—Individual prismatic crystals have a pseudohexagonal cross section; columnar aggregates are common.

Cleavage—One good cleavage is apparent in some orientations; an additional poor cleavage is rarely seen.

Relief—High.

Interference Colors—Clinozoisite displays anomalous or normal first-order colors; epidote displays lower second-order to upper third-order interference colors, depending on composition. Anomalous blue, green, or yellow colors are common for both.

Extinction and Orientation—Parallel extinction in longitudinal section; can be either length fast or length slow.

Twinning—Simple twins may be present.

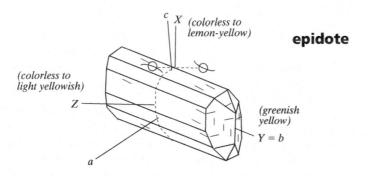

▶H. NATIVE ELEMENTS

Graphite   **C**

Plate—30e–f

Occurrence—A widespread accessory mineral in schists, marbles, gneisses, and other metamorphic rocks. It is a rare mineral in some igneous rocks.

Distinguishing Features—Opaque, metallic luster, and black under reflected light.

Similar Minerals—Can be confused with molybdenite or other silver or gray opaque materials.

Properties and Interference Figure—Hexagonal, opaque.

Color—Black with reflected light.

Form—Small flakes or scattered specks are typical; they may appear as thin tabs when viewed edge-on.

Cleavage—One good, rarely visible.

▶ I. SULFIDES

Sphalerite   ZnS

Occurrence—An important ore mineral, sphalerite is found in hydrothermal deposits and in some contact aureoles.

Distinguishing Features—May be opaque (brown-red to black in reflected light). If transparent, it is gray, yellow, or brown, shows complicated cleavage patterns, has high relief, and is isotropic.

Similar Minerals—If opaque, it may be hard to distinguish from other reddish brown opaque minerals. If transparent, it may be confused with rutile.

Properties and Interference Figure—Cubic; isotropic; $n = 2.39$–2.42.

Color—Normally yellow, yellow-brown or brown; more rarely gray or colorless.

Form—Irregular masses are common; euhedral crystals less common.

Cleavage—Six cleavages yield complicated patterns when viewed in thin sections.

Relief—Very high.

Pyrite   FeS$_2$

Plates—7e–f, 33e–f

Occurrence—An accessory mineral in many igneous, metamorphic and sedimentary rocks.

Distinguishing Features—An opaque, brass yellow color in reflected light; crystals are often square, triangular, or rectangular in cross sections.

Similar Minerals—May be difficult to tell from other sulfide minerals, especially chalcopyrite, marcasite, or pyrrhotite.

Properties and Interference Figure—Cubic; opaque.

Color—Brass yellow in reflected light.

Form—Euhedral crystals can be square, triangular, or rectangular.

Cleavage—One poor cleavage; subconchoidal fracture common.

Pyrrhotite $Fe_{1-x}S$

Occurrence—Typically found in mafic igneous rocks. Less commonly in hydrothermal veins or contact aureoles.

Distinguishing Features—Opaque, bronze to copper red color; typically in masses, blades, or as disseminated grains.

Similar Minerals—Pyrrhotite is somewhat similar to some other sulfides, but the bronze to reddish color may be diagnostic.

Properties and Interference Figure—Hexagonal, opaque.

Color—Bronze to copper red in reflected light.

Form—May be massive or bladed, or may be scattered as disseminated grains.

Cleavage—One poor cleavage, uneven fracture.

Chalcopyrite   CuFeS$_2$

Plate—34a–b

Occurrence—Found in most sulfide deposits, associated with hydrothermal veins or replacements.

Distinguishing Features—Opaque, bright brass color in reflected light; rare euhedral crystals, typically massive.

Similar Minerals—Similar to pyrite, but color is more yellow. Sometimes also confused with marcasite or pyrrhotite.

Properties and Interference Figure—Tetragonal; opaque.

Color—Bright brass yellow color (reflected light) is more yellow than pyrite.

Form—Normally massive, euhedral crystals are rare.

Cleavage—One poor cleavage; fractures common.

▶J. HALIDES

Fluorite CaF_2

Occurrence—Found in veins and carbonate-hosted ore deposits, as an accessory mineral in limestone or dolomite, and in igneous and metamorphic rocks.

Distinguishing Features—Clear to light purple; high relief; perfect octahedral cleavages; isotropic.

Similar Minerals—Opal has no cleavage. Sodalite has greater relief and different cleavage.

Properties and Interference Figure—Cubic; isotropic; $n = 1.434$.

Color—Colorless to light purple or blue; color may be zoned; "growth rings" are common; pleochroic halos may be present.

Form—Usually anhedral in patches or vein fillings; rare crystals are euhedral with square cross sections.

Cleavage—Four perfect cleavages, two or three are visible depending on orientation.

Relief—Fairly high.

Twinning—Penetration twins are common but rarely visible.

▶K. OXIDES

Rutile   **TiO₂**

Plate—18e–f

Occurrence—Typically found as small grains in intermediate to mafic igneous rocks, in some metamorphic rocks, in veins, in pegmatites, and in some sediments.

Distinguishing Features—Strong yellowish to reddish brown color is typical but fine needles may be pale (Plate 18e–f); very high relief; extreme birefringence; adamantine luster. Typically small grains.

Similar Minerals—Rutile's red color may sometimes resemble the color of (thin) hematite. It may also be confused with cassiterite (lower relief and birefringence), lepidocrocite, goethite, and brookite.

Properties and Interference Figure—Tetragonal; uniaxial (+); $\omega = 2.609$–2.616, $\varepsilon = 2.895$–2.903, $\delta = 0.286$–0.287. Interference figures show many isochromes.

Color—Yellowish to reddish brown in thin sections (ω = yellow to red brown, ε = brownish to yellowish green); adamantine luster in reflected light.

Form—Small grains and prismatic or needle-like crystals; simple twins forming "kinks" are common.

Cleavage—Two good cleavages, but rarely visible.

Relief—Very high.

Interference Colors—Extreme birefringence; interference colors are extreme but rarely show due to coloration and light reflection.

Extinction and Orientation—Parallel extinction.

Twinning—Simple twinning is very common.

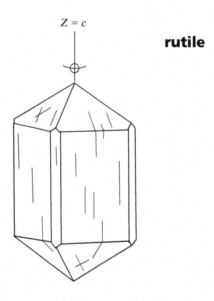

rutile

Hematite Fe_2O_3

Plates—7a–b, 9c–d, 10e–f, 24c–d, 27c–d, 34e–f

Occurrence—A common mineral in many kinds of rocks, including most notably red sandstone. It can form as an alteration product after just about any Fe-bearing mineral.

Distinguishing Features—Normally opaque, but thin crystals may show some red color at their edges (most noticeable with the conoscopic lens inserted); red, black, or steel blue in reflected light.

Similar Minerals—Similar to goethite and lepidocrocite, but they are lighter colored and more brownish. Ilmenite is darker and blacker.

Properties and Interference Figure—Hexagonal (rhombohedral); uniaxial (−); ω = 2.90–3.22, ε = 2.69–2.94, δ =0.21–0.28.

Color—Normally opaque; red, black, or steel blue in reflected light; thin edges of otherwise opaque grains in thin sections may appear deep red. If transparent, appears brownish red to yellowish red in transmitted light.

Form—Euhedral crystals and masses are both common.

Cleavage—No good cleavages, but often fractured.

Relief—Very high.

Twinning—Common but not often visible in thin sections.

Corundum   Al$_2$O$_3$

Plate—32c–d

Occurrence—An accessory mineral in metamorphosed carbonates, in some other aluminous metasediments, in some Al-rich igneous rocks, and in placers. Massive corundum is found in skarns.

Distinguishing Features—High relief, low birefringence, parting, sometimes skeleton crystals (crystals full of holes) and, when present, twinning.

Similar Minerals—Vesuvianite and apatite have less birefringence and relief.

Properties and Interference Figure—Hexagonal (rhombohedral); uniaxial (−); ω = 1.767–1.772, ε = 1.759–1.763, δ = 0.005–0.009. Rare corundum is biaxial.

Color—Typically colorless; sometime brownish or showing patchy areas of blue (sapphire) or pink (ruby); zoned crystals are common; may be pleochroic (ω = blue-purple, ε = light to greenish yellow).

Form—Euhedral hexagonal or tabular crystals are common and often zoned; longitudinal sections are prismatic.

Cleavage—Good parting in several directions but no good cleavages; basal sections may show three partings at 60° to each other (Plate 32c–d).

Relief—Very high.

Interference Colors—Low birefringence, but corundum may be overly thick in thin sections because it is hard to grind, so interference colors range up to second-order.

Extinction and Orientation—In basal sections, extinction is symmetrical to parting; in longitudinal sections, it is parallel; prismatic crystals are length fast and tabular crystals are length slow.

Twinning—Twins are common, often forming irregular lamellae.

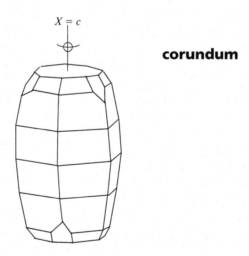

corundum

$X = c$

Spinel $(Mg,Fe,Zn)Al_2O_4$

Plate—31a–d

Occurrence—A high temperature mineral found in marbles, metamorphosed marls, schists, or gneisses; also found as an accessory in mafic igneous rocks.

Distinguishing Features—Colorless to green, less commonly light red, brown, or other colors; isotropic; high relief; octahedral form.

Similar Minerals—Garnet and spinel may be difficult to distinguish, but garnet is typically lighter colored and poikiloblastic. Spinel is sometimes difficult to distinguish from periclase. Perovskite may resemble spinel but has a much higher refractive index.

Properties and Interference Figure—Cubic; isotropic; $n = 1.72$–1.74.

Color—End member spinel is colorless but becomes green (pleonaste) if Fe is present; hercynite is end member $FeAl_2O_4$. Pink, bluish, olive green, and brown varieties are also known.

Form—Euhedral to subhedral crystals often form equant grains and may display rhombic sections of octahedra.

Cleavage—None, but often fractured.

Twinning—Commonly twinned.

Ilmenite   FeTiO₃

Plate—34c–d

Occurrence—A common vein mineral, also found in many different kinds of igneous rocks including pegmatites. It may also be an accessory mineral in metamorphic rocks.

Distinguishing Features—Opaque, violet black in reflected light, flake-like crystals, associated with magnetite and alters to a cottony opaque white substance called leucoxene.

Similar Minerals—More violet or purple than red hematite; may be difficult to tell from magnetite unless both are present in the same thin section.

Properties and Interference Figure—Hexagonal (rhombohedral), opaque.

Color—Violet black in reflected light; more violet or purple than hematite.

Form—Disseminated tabular or flaky crystals, skeletal crystals, irregular grains, or masses.

Cleavage—None, but often fractured.

Twinning—Commonly twinned.

Magnetite

Fe₃O₄

Plates—12c–d, 25a–b, 31a–b, 34c–d

Occurrence—Common and widespread in igneous, metamorphic, and sedimentary rocks.

Distinguishing Features—Opaque, steel blue-black in thin sections, euhedral crystals are rhombohedral cross sections of octahedra and are magnetic.

Similar Minerals—Ilmenite tends to be more purple or violet (reflected light) than magnetite (Plate 34c–d); chromite is more iron black or brown black.

Properties and Interference Figure—Cubic, opaque; magnetic.

Color—Opaque, steel blue-black in reflected light.

Form—Euhedral crystals are rhombohedral cross sections of octahedra; massive or granular aggregates are also common.

Cleavage—None; fracture is common.

Twinning—Contact or lamellar twins are common.

Chromite   FeCr$_2$O$_4$

Plate—19a–b

Occurrence—Primarily found in ultramafic rocks, typically associated with olivine.

Distinguishing Features—Opaque; iron black to brownish black; may appear brown at the edges in focused transmitted light.

Similar Minerals—May be mistaken for graphite, black hematite, ilmenite, or magnetite, but chromite normally has a more steel-like color in reflected light.

Properties and Interference Figure—Cubic; isotropic, nearly opaque; n = 2.05–2.16.

Color—Iron black to brownish black in reflected light; opaque or nearly opaque in transmitted light.

Form—Euhedral crystals are rare; normally massive or granular aggregates.

Cleavage—None.

▶L. HYDROXIDES

Gibbsite $Al(OH)_3$

Occurrence—A secondary mineral associated with aluminous silicates.

Distinguishing Features—Fine-grained, clear to light brown, high first-order or second-order colors, and occurs on cavity walls or as an alteration product.

Similar Minerals—Gibbsite resembles chalcedony but has greater relief and much greater birefringence. Brucite is uniaxial. Kaolinite and lepidolite are more micaceous; kaolinite has lower birefringence and lepidolite is biaxial $(-)$.

Properties and Interference Figure—Monoclinic, biaxial $(+)$; $2V = 0°$ to $40°$; $\alpha = 1.567$, $\beta = 1.567$, $\gamma = 1.588$, $\delta = 0.021$. Crystals are typically too small to reveal interference figures.

Color—Colorless to pale brown.

Form—Individual crystals are normally very small pseudohexagonal prisms; masses and fine grained aggregates are typical.

Cleavage—One perfect cleavage, but often hard to see.

Relief—Moderate.

Interference Colors—Maximum colors are first-order red or low second-order.

Extinction and Orientation—Elongate sections have an extinction angle up to about $27°$, cleavage fragments, and are length fast.

Twinning—Polysynthetic twinning is common.

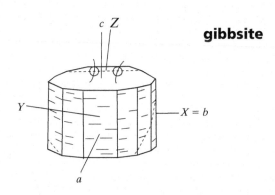

gibbsite

▶M. CARBONATES

Calcite   CaCO₃

Plates—3c–d, 8a–f, 10a–d, 14e–f, 20e–f, 26a–b, 33c–f

Occurrence—Common in limestone, dolomite, marbles, marls, hydrothermal veins, and some rare igneous rocks. It is also a secondary mineral in rocks of many different types.

Distinguishing Features—Colorless, extreme birefringence, variable relief with stage rotation, rhombohedral cleavage, and commonly twinned. A red stain that only affects calcite is often applied to thin sections to distinguish it from dolomite (Plates 20e–f, 33e–f).

Similar Minerals—Similar in many ways to other rhombohedral carbonates including dolomite, magnesite, and siderite. Dolomite tends to have twins parallel to both the short and long diagonals, magnesite shows no twins, and siderite often has hematite stains. Distinguishing the various carbonates can be difficult in thin sections. Aragonite, calcite's orthorhombic polymorph, is biaxial and does not show rhombohedral cleavage.

Properties and Interference Figure—Hexagonal (rhombohedral); uniaxial (−); ω = 1.658, ε = 1.486, δ = 0.172. Very rare calcite is biaxial.

Color—Colorless, but may be cloudy or show vague pastel hues or twinkling effects.

Form—Euhedral crystals are rare in thin sections; fine to coarse-grained anhedral aggregates are typical; they may show relict organic structures (fossils, Plates 8a–b, 10c–d).

Cleavage—Calcite has perfect rhombohedral cleavage; if coarse enough, most sections show two intersecting cleavages.

Relief—Most orientations show variable relief (high to low) on stage rotation.

Interference Colors—Very high order, often appearing as white; edges of grains may show lower (fourth- or fifth-order) color.

Extinction and Orientation—Extinction is symmetrical to cleavages.

Twinning—Polysynthetic twinning is very common and usually visible, sometimes even in PP light; twin lamellae, parallel to the long rhomb diagonal, are often thin enough to show first-order colors.

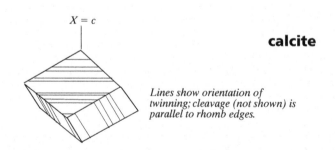

X = c

calcite

Lines show orientation of twinning; cleavage (not shown) is parallel to rhomb edges.

Magnesite MgCO₃

Magnesite $MgCO_3$

Plate—25a–b

Occurrence—Occurs in low- to medium-grade metamorphic rocks and in ultramafic rock, most commonly as an alteration product of mafic minerals. It also occurs in Mg-rich schists, in some rare chemical sediments, and as an alteration product in limestones and dolomites.

Distinguishing Features—Colorless; extreme birefringence; variable relief with stage rotation; no twins.

Similar Minerals—Difficult to distinguish from calcite and other carbonates (see calcite on page 104).

Properties and Interference Figure—Hexagonal (rhombohedral); uniaxial ($-$); $\omega = 1.700, \varepsilon = 1.509, \delta = 0.191$.

Color—Colorless.

Form—Anhedral to subhedral crystal aggregates.

Cleavage—Perfect rhombohedral cleavage; most sections show two cleavages if grains are large enough.

Relief—Varies from low to high with stage rotation.

Interference Colors—Extremely high-order white/pearl.

Extinction and Orientation—Symmetrical extinction with respect to cleavages.

Twinning—Absent.

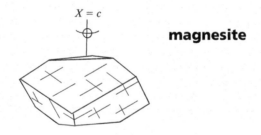

magnesite

Siderite   **FeCO₃**

Plate—10e–f

Occurrence—Found in hydrothermal veins, in limestones or dolomites, as a replacement mineral, and less commonly in metamorphic rocks.

Distinguishing Features—Colorless but often has patchy yellow or brown staining, variable relief on rotation, and extreme birefringence.

Similar Minerals—Resembles other carbonates, but Fe staining helps distinguish it. Other carbonates normally have lower relief, but this is only a useful property if they are in the same thin section. Cassiterite has lower birefringence and is optically (+). Titanite is biaxial.

Properties and Interference Figure—Hexagonal (rhombohedral); uniaxial (−); $\omega = 1.851$–1.875, $\varepsilon = 1.612$–1.633, $\delta = 0.239$–0.242.

Color—Colorless to gray, often stained yellow or brown by hematite and other Fe compounds.

Form—Typically in aggregates, fine- to coarse-grained; euhedral crystals rare.

Cleavage—Perfect rhombohedral cleavage; most sections show two cleavage traces if coarse enough.

Relief—Varies from moderate to high with stage rotation.

Interference Colors—White (pearl) of very high-order.

Extinction and Orientation—Symmetrical extinction with respect to cleavage.

Twinning—Twins are sometimes present.

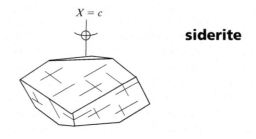

siderite

Dolomite CaMg(CO₃)₂

Dolomite $CaMg(CO_3)_2$

Plate—8e–f

Occurrence—Common in dolomites and in marbles; it also occurs in hydrothermal veins, as a secondary mineral, and as an alteration product in limestones.

Distinguishing Features—Colorless to gray; extreme birefringence; variable relief with stage rotation, and commonly twinned. A red stain can be applied to thin sections to distinguish calcite (which stains) from dolomite (which does not).

Similar Minerals—Similar in many ways to calcite, but calcite has lower relief, tends to form only subhedral to euhedral crystals, is often zoned, and does not have twin lamellae parallel to the short diagonal. Dolomite is more similar to magnesite (which does not twin); if dolomite is untwinned, distinguishing the two is very difficult in thin section.

Properties and Interference Figure—Hexagonal (rhombohedral); uniaxial (−); $\omega = 1.679$–1.698, $\varepsilon = 1.502$–1.513, $\delta = 0.177$–0.185. Interference figures show many isochromes.

Color—Colorless; rarely gray or brownish.

Form—Usually subhedral, but euhedral crystals are common; crystals may be bent or curved; fine- to coarse-grained anhedral aggregates are common; zoning is often visible due to variable Fe content.

Cleavage—Dolomite has perfect rhombohedral cleavage; if coarse enough, most sections show two intersecting cleavages.

Relief—Most orientations show variable relief (high to low) on stage rotation.

Interference Colors—Very high order, often appearing as white; edges of grains may show lower (fourth- or fifth-order) color.

Extinction and Orientation—Extinction is symmetrical to cleavages; curved crystals show wavy extinction.

Twinning—Polysynthetic twinning is very common and usually visible; twin lamellae, parallel to the long or short rhomb diagonals, are often thin enough to show first-order colors.

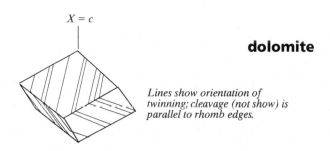

$X = c$

dolomite

Lines show orientation of twinning; cleavage (not show) is parallel to rhomb edges.

▶N. SULFATES

Anhydrite   CaSO$_4$

Plate—11a–b

Occurrence—Typically found in evaporite or metaevaporite deposits; also found in amygdules or cracks in basalt, as a minor mineral in hydrothermal deposits, and in hot spring deposits.

Distinguishing Features—Colorless, third-order interference colors; rectangular cleavage.

Similar Minerals—Higher relief and stronger birefringence than gypsum.

Properties and Interference Figure—Orthorhombic; biaxial (+); 2V = 42°; α = 1.570, β = 1.576, γ = 1.614, δ = 0.044.

Color—Colorless.

Form—Typically, anhedral to subhedral crystals form fine- to medium-grained aggregates.

Cleavage—Three cleavages at 90°; two are visible in most sections.

Relief—Moderate; small variation with stage rotation.

Interference Colors—Maximum colors are third-order green.

Extinction and Orientation—Extinction is parallel to cleavages.

Twinning—Polysynthetic twinning is common.

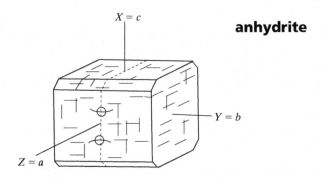

Barite $BaSO_4$

Plates—6d–g, 11e–f

Occurrence—A common minor mineral in hydrothermal veins; also found in veins in limestone, and as residual masses in clays.

Distinguishing Features—Colorless, high relief, low birefringence, and 90° cleavage angles.

Similar Minerals—Hard to tell from celestite, another sulfate, although celestite has a greater *2V*.

Properties and Interference Figure—Orthorhombic; biaxial (+); $2V = 26°$ to $38°$; $\alpha = 1.636$, $\beta = 1.637$, $\gamma = 1.648$, $\delta = 0.012$.

Color—Colorless.

Form—Rare individual crystals are elongate orthorhombic prisms; granular aggregates are typical.

Cleavage—Four good-perfect cleavages; the principal ones intersect at 90°; two are visible in most sections.

Relief—High.

Interference Colors—Low birefringence results in maximum colors being first-order yellow or orange.

Extinction and Orientation—Extinction is parallel to the best cleavage, but not to all cleavages.

Twinning—Rare polysynthetic twinning.

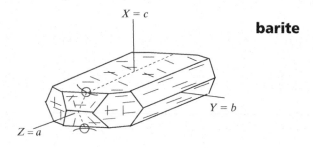

barite

Gypsum $CaSO_4 \cdot 2H_2O$

Plate—11c–d

Occurrence—Massive gypsum is associated with evaporites; it is also found interlayered in limestones and shales, as fillings in fractures or holes in many sedimentary rocks, and as a gangue mineral or alteration product in some ore deposits.

Distinguishing Features—Colorless, low relief, low interference colors, and 66° cleavage angle.

Similar Minerals—Lower relief and birefringence than anhydrite. Brucite has higher relief.

Properties and Interference Figure—Monoclinic; biaxial (+); $2V = 58°$; $\alpha = 1.520$, $\beta = 1.523$, $\gamma = 1.530$, $\delta = 0.010$.

Color—Colorless.

Form—Anhedral to subhedral aggregates, often with variable grain size; rarely fibrous.

Cleavage—Several good-perfect cleavages.

Relief—Low.

Interference Colors—Maximum colors are first-order weak yellow.

Extinction and Orientation—Extinction is parallel to the best cleavage.

Twinning—Polysynthetic twinning is often induced during thin section-making

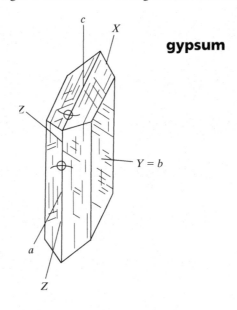

gypsum

▶ O. PHOSPHATES

Apatite $Ca_5(PO_4)_3(F,Cl,OH)$

Plates—10a–b, 17a–b, 18e–f, 33a–b

Occurrence—A common accessory mineral in many different igneous and metamorphic rocks, but often overlooked.

Distinguishing Features—Colorless, moderate relief; white-to-gray interference colors; often small lathlike prismatic crystals with a hexagonal cross section; uniaxial (−).

Similar Minerals—When coarse, it is similar to quartz in many ways, but quartz is optically (+) and has lower relief. Zoisite has higher relief. Nepheline and beryl have lower relief. Collophane, a fine-grained variety of apatite is found in some sedimentary rocks (Plate 10a–b).

Properties and Interference Figure—Hexagonal; uniaxial (−); $\omega = 1.634–1.651$, $\varepsilon = 1.631–1.646$, $\delta = 0.019–0.05$. Basal (tabular) sections are usually too small to observe interference figures.

Color—Colorless, less commonly brownish or reddish; inclusions may give a gray or black appearance.

Form—Sometimes coarse anhedral crystals, but more typically small subhedral to euhedral hexagonal crystals or prisms.

Cleavage—Several poor cleavages in one direction, but rarely seen in thin sections.

Relief—Moderate.

Interference Colors—Maximum colors are white and gray.

Extinction and Orientation—Parallel extinction in long sections; longitudinal (prismatic) sections are length fast, but basal (semi-hexagonal) sections are length slow.

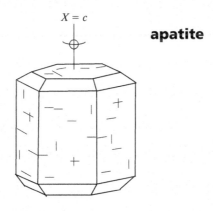

PLATE 7: MINERALS IN SEDIMENTARY ROCKS

Quartz, Hematite, Chalcedony, clay, Pyrite

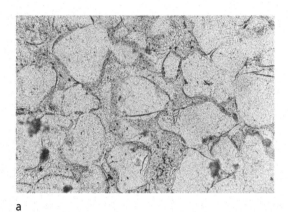

a

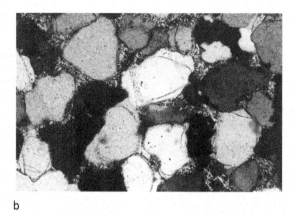

b

Photographs a (PP, 40x) and b (XP, 40x): A red sandstone from Potsdam, New York. Most of the grains in the photograph are <u>Quartz</u> with rings of <u>Hematite</u> within the grains. The rings mark the surfaces of the original sand grains. After burial, some <u>Quartz</u> partially dissolved in groundwater and reprecipitated over the <u>Hematite</u>-coated <u>Quartz</u> grains, causing the <u>Hematite</u> rings to now appear within the grains. The finer-grained material between the grains is probably <u>Chalcedony</u>.

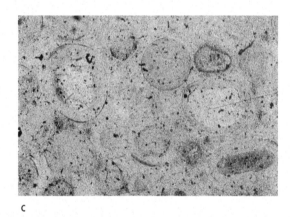

c

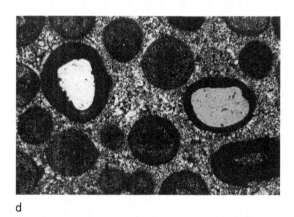

d

Photographs c (PP, 40x) and d (XP, 40x): A siliceous oolite from State College, Center County, Pennsylvania. The rock was originally an oolitic limestone. <u>Quartz</u> grains were rolled into limestone pellets (oolites), by near-shore waves. After burial and over time, the round oolites and their surrounding finer-grained <u>Calcite</u> matrix were replaced by Silica (mostly <u>Chalcedony</u>) from groundwater. Two <u>Quartz</u> grains are visible in the centers of two oolites.

e

f

Photographs e (PP, 40x) and f (XP, 40x) of an oil shale from Garfield County, Colorado. The shale consists of <u>Clay</u>, organic-rich materials, <u>Pyrite</u>, and <u>Quartz</u>. Most Shales, like this one, are too fine-grained to show a lot of interesting features in a thin section. However, <u>Quartz</u> and brown organic-rich materials are noticeable in this thin section.

112

Calcite, Quartz, Dolomite

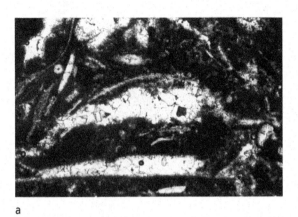 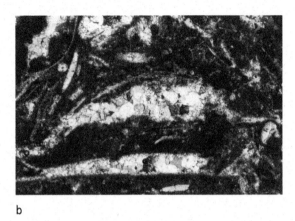

a b

Photographs a (PP, 40x) and b (XP, 40x) of a fossiliferous limestone from an unknown locality. The elliptical feature in the center of the photograph is a cross-section of a brachiopod shell in a generally fine-grained, dark, Calcite-rich matrix. The shell is filled by coarser-grained Calcite. Notice that the birefringence of Calcite is so high (pearl white interference colors) that it is difficult to tell whether the Calcite is in PP or XP light.

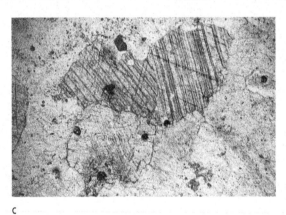 

c d

Photographs c (PP, 40x) and b (XP, 40x) of three twinned Calcite grains in a graywacke from near Grafton, New York. Some of the twins are off-set by rhombohedral fractures, as seen in the grain in the upper-right part of the XP photograph. The three major Calcite grains are surrounded mostly by Quartz, some of which is very fine-grained, as shown at the upper-left edge of the photographs.

e f

Photographs e (PP, 100x) and f (XP, 100x) of Dolomite in a limestone from Lockport, New York. Rhombs of Dolomite are most easily seen towards the center of the XP photograph. Fine-grained Calcite and Dolomite both have high interference colors and may be difficult to distinguish in thin section. Sometimes a red solution is applied to thin sections to distinguish Calcite from Dolomite and other minerals. The solution stains Calcite red (see, for examples, photograph e in Plate 20 and photograph e in Plate 33), but does not affect Dolomite and most silicates.

PLATE 9: MINERALS IN SEDIMENTARY ROCKS (CONTINUED)

Chalcedony, Quartz, Muscovite, K-feldspar, Hematite, Glauconite, Chert

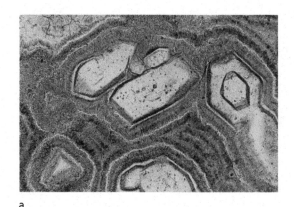

a

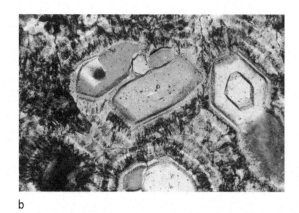

b

Photographs a (PP, 100x) and b (XP, 100x) of chalcedony surrounding euhedral to subhedral quartz grains in a concretion from Burnside, Kentucky. Many <u>Quartz</u> grains show first-order yellow interference colors because the thin section is a bit too thick. The feathery texture between quartz grains developed as the <u>Chalcedony</u> crystallized, growing out from the quartz grains.

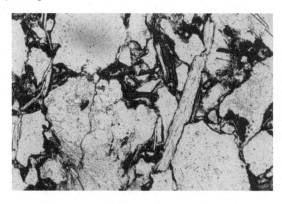

c

d

Photographs c (PP, 100x) and d (XP, 100x) of an arkose (Sugarloaf Fm., Mt. Tom, Massachusetts, Triassic). The major minerals are <u>K-feldspar</u> and <u>Quartz</u>, both of which are clear (PP) and display low order interference colors (XP). They are difficult to tell apart in this photograph. Long skinny flakes of <u>Muscovite</u> display high order interference colors. <u>Hematite</u> (dark reddish brown) stains some grains and fills some interstices.

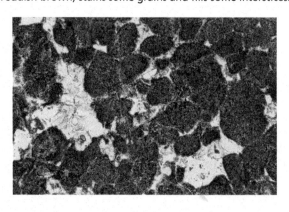

e

f

Photographs e (PP, 100x) and f (XP, 100x) of a glauconitic sandstone (Borden Fm., Bighill, Kentucky, Mississippian). <u>Glauconite</u> (green mineral), an Fe-rich mica-like mineral that is often grouped with clays, is often poorly crystalline or amorphous. The clear, low birefringence material between glauconite grains is <u>Chert</u>. Most glauconitic sandstones contain less glauconite than this specimen. This section is from the collection of Frank Ettensohn, University of Kentucky.

Calcite, Quartz, Collophane, Siderite, Hematite

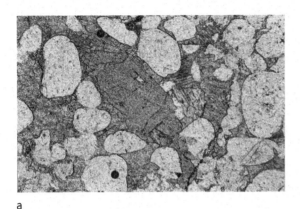
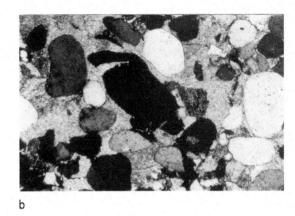

a

b

Photographs a (PP, 100x) and b (XP, 100x) of a sandstone with calcite cement. Quartz grains are well-rounded and display only first-order white to gray interference colors; the Calcite is fine-grained and displays high-order interference colors. The large brown, isotropic, grain in the center of the photograph is most likely Collophane, a fine-grained variety of Apatite. This sample is from the collection of Frank Ettensohn, University of Kentucky.

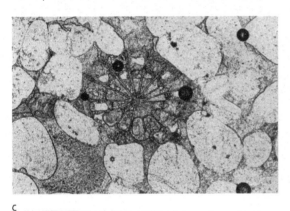
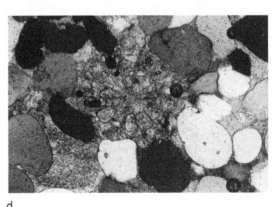

c

d

Photographs c (XP, 100x) and d (PP, 100x) of a sandstone. Here we see well-rounded Quartz grains and a fine-grained Calcite cement. The large geometric pattern at the center of the photo is a fossil fragment made of Calcite. This sample comes from an unknown location.

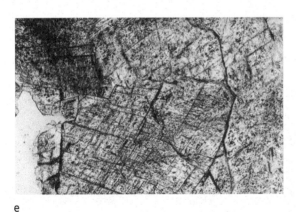

e

f

Photograph e (PP, 100x) and f (XP, 100x) of siderite in a carbonate geode from Burnside, Kentucky. The Siderite shows well-developed cleavage, typical of Siderite, Calcite and other rhombohedral carbonate minerals. It is lightly stained by Hematite. Siderite is the only mineral present; the clear, isotropic, area in the corner of the photograph is a hole in the thin section.

Anhydrite, Gypsum, Barite

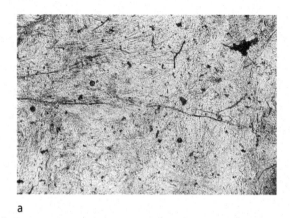
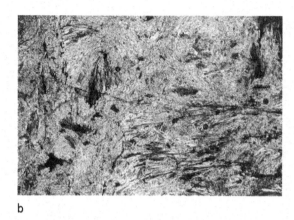

a

b

Photographs a (PP, 40x) and b (XP, 40x) of Anhydrite (CaSO₄) from Hants County, Nova Scotia. <u>Anyhdrite</u>'s birefringence is about 0.044, so third order interference colors are seen.

c

d

Photographs c (PP, 40x) and d (XP, 40x) of Gypsum (CaSO₄ • 2H₂O) from Grand Rapids, Michigan. <u>Gypsum</u> has lower interference colors than <u>Anhydrite</u>. In PP light, <u>Gypsum</u> shows few details. However, a large elongated grain of <u>Gypsum</u> is clearly visible in XP light. It is surrounded by finer grains of <u>Gypsum</u>.

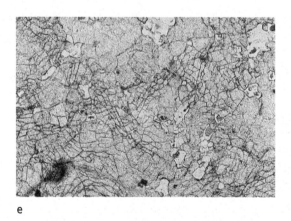

e

f

Photographs e (PP, 40x) and f (XP, 40x) of Barite from King's Creek, South Carolina. The field of view consists almost entirely of <u>Barite</u>. Note the near 90° cleavage angle.

PLATE 12: MINERALS IN IGNEOUS ROCKS

Glass, opaques, Plagioclase, Hornblende, Clinopyroxene, Biotite, Quartz, Magnetite, Pyroxene

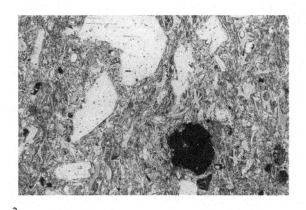
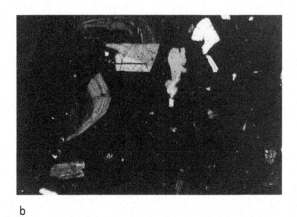

a

b

Photographs a (40x, PP) and b (40x, XP) of a volcanic tuff from an unknown location. The sample contains abundant fibrous glass, opaque minerals, and zoned and twinned Plagioclase.

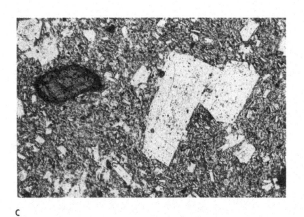

c

d

Photographs c (PP, 40x) and d (XP, 40x) of hornblende andesite from Mt. Shasta, California. Zoned Plagioclase with twins appears in a finer-grained matrix of volcanic glass, Plagioclase, Clinopyroxene and Magnetite. The concentric zoning and Carlsbad twinning in the Plagioclase are easily seen in XP light. On the left, a brown Hornblende with a reaction rim of fine-grained materials (perhaps Magnetite and Pyroxene) can be seen.

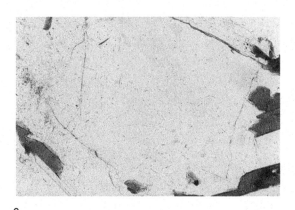

e

f

Photographs e (PP, 100x) and f (XP, 100x) of undulatory Quartz surrounded by Biotite in a biotite granite from Barre, Vermont. Tectonic or other forces have slightly deformed the Quartz grain so that different parts of the grain go extinct at different times. In PP light, the Biotite appears brown.

PLATE 13: MINERALS IN INTRUSIVE IGNEOUS ROCKS

Microcline, Plagioclase, Biotite, Nepheline, Augite

a

b

Photographs a (PP, 40x) and b (XP, 40x) of Microcline in a muscovite biotite granite from Concord, New Hampshire. In XP light, the <u>Microcline</u> has visible Scotch plaid twinning. The <u>Microcline</u> is surrounded by <u>Biotite</u> (yellow to brown in PP light) and <u>Plagioclase</u>.

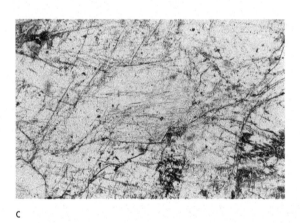

c

d

Photographs c (PP, 40x) and d (XP, 40x) of sodic Plagioclase in a nepheline syenite from Blue Mountain, Methuen Township, Ontario. The <u>Plagioclase</u> has lamellar (albite) twinning and is surrounded by <u>Nepheline</u>.

e

f

Photographs e (PP, 40x) and f (XP, 40x) of calcic Plagioclase and Augite in an olivine gabbro from Wichita Mountain, Oklahoma. The <u>Plagioclase</u> displays Pericline and Albite twins, visible in XP light.

PLATE 14: MINERALS IN INTRUSIVE IGNEOUS ROCKS (CONTINUED)

Plagioclase, Sericite, Orthoclase, Hornblende, opaques, Biotite, Calcite

a

b

Photographs a (PP, 40x) and b (XP, 40x) of a zoned Plagioclase in a quartz monzonite porphyry from Garfield, Colorado. The concentric zoning, which results from variations in composition of the <u>Plagioclase</u>, is easily seen in XP light. Minor veins and patches of <u>Sericite</u> (fine-grained white <u>Micas</u> produced by weathering or metamorphic alteration of the <u>Plagioclase</u>) cover parts of the <u>Plagioclase</u>.

c

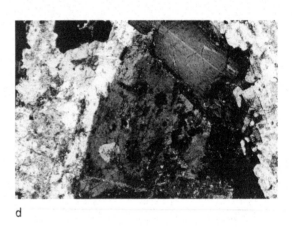

d

Photographs c (PP, 100x) and d (XP, 100x) of sericitized Orthoclase in a hornblende syenite from Cuttingsville, Vermont. In PP light, the <u>Sericite</u> appears as a brown dust covering the <u>Orthoclase</u>. A brown <u>Hornblende</u> is located in the upper-right corner of the photographs. In the top left, <u>Opaques</u> and a brown <u>Biotite</u> are present.

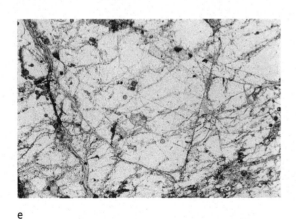

e

f

Photographs e (PP, 40x) and f (XP, 40x) of Orthoclase from Good Springs, Nevada. Veins of <u>Sericite</u> and <u>Calcite</u> are present in parts of the thin section.

PLATE 15: MINERALS IN IGNEOUS ROCKS

Anorthoclase, Olivine, Nepheline, Sericite, Leucite, Plagioclase, opaques

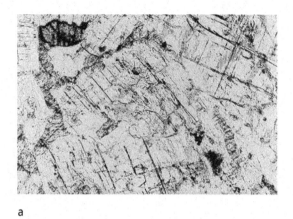

a

b

Photographs a (PP, 40x) and b (XP, 40x) of Anorthoclase grains in a sample from Larvik, Norway. A small grain of high-relief <u>Olivine</u> is located in the upper-left corner of the photographs. It appears nearly extinct in XP light.

c

d

Photographs c (PP, 40x) and d (XP, 40x) of Nepheline in a nepheline syenite from Blue Mountain, Ontario. <u>Nepheline</u> may resemble Quartz, except that it is uniaxial negative. Veins of <u>Sericite</u> cross the sample and are visible under XP light.

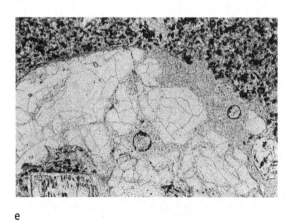

e

f

Photographs e (PP, 40x) and f (XP, 40x) of Leucite in a vesicle in a sample from Rome, Italy. The <u>Leucite</u> is almost isotropic and so appears nearly black in XP light. It does, however, show laminar twinning. Isotropic epoxy surrounds some of the Leucite grains. Two air bubbles are also present between the <u>Leucite</u> grains, and their "eclipse halos" may be seen in the XP photograph. The matrix along the top portion of the photographs contains abundant fine-grained <u>Opaques</u> and <u>Plagioclase</u>. Several grains of <u>Plagioclase</u> appear white in the XP photograph.

PLATE 16: COMMON MINERAL TEXTURES IN GRANITIC ROCKS

Perthite, Myrmekite, Quartz, Plagioclase, K-feldspar

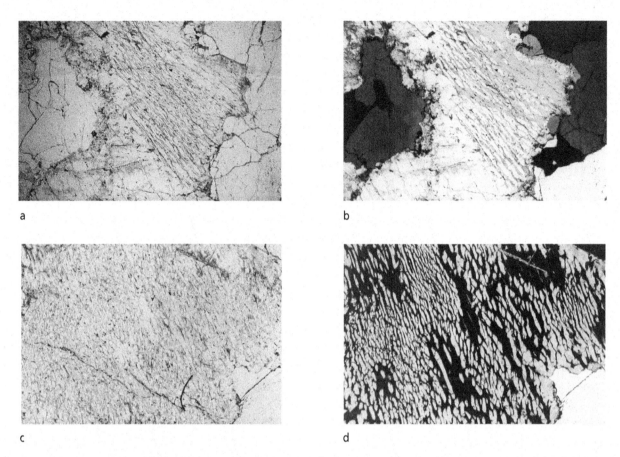

a b

c d

Photographs a through d (40x) of Perthite in a migmatite from near Ear Falls, Ontario. Perthite consists of patches or lenses of sodic Plagioclase in a K-feldspar matrix. The two feldspars formed a solid solution at high temperatures but separated as temperatures fell. Photographs a (PP) and b (XP) show a partially perthitized Feldspar with Quartz grains on the left and right. Photographs c (PP) and d (XP) show close-ups of Perthite and have a Quartz grain in the lower-right corner.

e f

Photographs e (PP, 100x) and f (XP, 100x) of Myrmekite in a granitic orthogneiss from near Vermillion Bay, Ontario. Myrmekite consists of "worm-like" inclusions of Quartz within Plagioclase. Quartz and Plagioclase surround the Myrmekite grain.

PLATE 17: MINOR MINERALS IN IGNEOUS ROCKS

Apatite, Plagioclase, Titanite, Quartz, Chlorite, Biotite, Microline, Hornblende

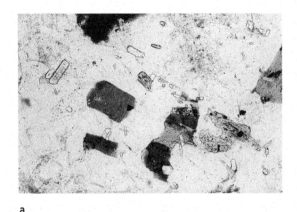

a

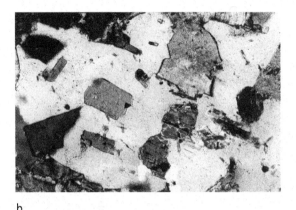

b

Photographs a (PP, 100x) and b (XP, 100x) of small prismatic grains of Apatite in a granodiorite from St. Cloud, Minnesota. In PP light, the <u>Apatite</u> appears clear and has high relief. In XP light, <u>Apatite</u> has low (gray) interference colors. The photographs also contain <u>Plagioclase</u>, <u>Quartz</u>, green <u>Chlorite</u>, and yellowish-brown <u>Biotite</u>.

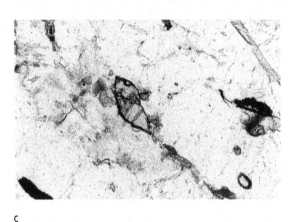

c

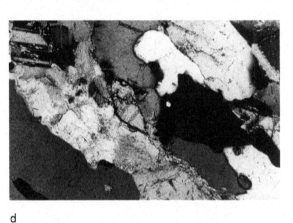

d

Photographs c (PP, 100x) and d (XP, 100x) of a rhombohedral Titanite (sphene) grain in an orthogneiss from near Otter Lake, Quebec. The high-relief, light-brown <u>Titanite</u> is surrounded by <u>Plagioclase</u>, <u>Quartz</u>, and small amounts of brown <u>Biotite</u> and green <u>Hornblende</u>. The three high relief, circular features in the lower-right corner of the photographs are probably air bubbles that were trapped under the cover slip of the thin section. Note the extremely high order interference colors displayed by <u>Titanite</u>.

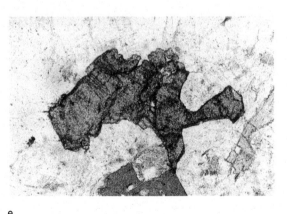

e

f

Photographs e (PP, 100x) and f (XP, 100x) of anhedral Titanite (sphene) in a granodiorite from St. Cloud, Minnesota. A grain of green <u>Hornblende</u> (at extinction in XP light) is located below the brown <u>Titanite</u>. Other surrounding grains include <u>Microcline</u>, <u>Plagioclase</u>, <u>Quartz</u> and a very pale green (in PP light) elongate <u>Hornblende</u> on the lower-right edge of the photographs.

122

Tourmaline, Biotite, Zircon, Quartz, opaques, Rutile, Muscovite, Apatite

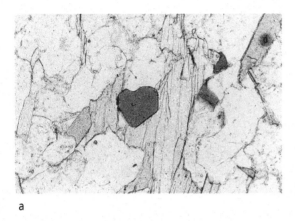 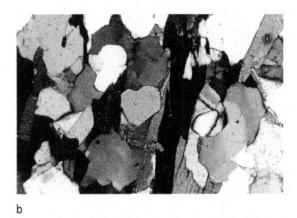

a   b

Photographs a (PP, 100x) and b (XP, 100x) of a green heart-shaped Tourmaline in a gneiss from Pakwash Lake, western Ontario. The green <u>Tourmaline</u> is surrounded by <u>Quartz</u> and light yellow to reddish-brown <u>Biotite</u>. A relatively large <u>Zircon</u> surrounded by a pleochroic halo is included in a biotite grain in the upper-right-hand corner of the photographs. The pleochroic halo resulted from radiation damage caused by radioactive decay of uranium within the <u>Zircon</u>.

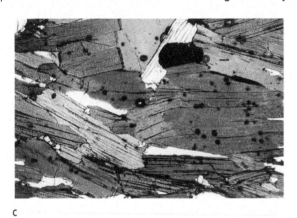 

c   d

Photographs c (PP, 40x) and d (XP, 40x) of Biotites, opaques and Zircons in a granulite from near Manitou Falls, western Ontario. In PP light, the <u>Biotites</u> are mostly greenish-brown to brown. Small <u>Zircon</u> inclusions and their pleochroic halos are scattered through the thin section and are best seen in PP light.

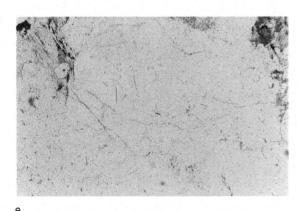 

e   f

Photographs e (PP, 200x) and f (XP, 200x) of many hair-like Rutile needles in Quartz in a biotite granite from Barre, Vermont. The <u>Rutile</u> needles are very small and faint. Grains of <u>Muscovite</u> and brown <u>Biotite</u> are located in the upper-left and right corners of the PP photograph. A hexagonal <u>Apatite</u> grain is associated with the <u>Muscovite</u> in the upper-left corner.

PLATE 19: OLIVINE AND PYROXENES

Olivine, Chlorite, Chromite, Iddingsite, Plagioclase, opaques, Augite

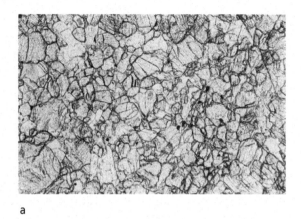

a

b

Photographs a (PP, 40x) and b (XP) of a dunite from near Balsam, North Carolina. The rock consists almost entirely of Olivine with small amounts of Chlorite and opaques (Chromite).

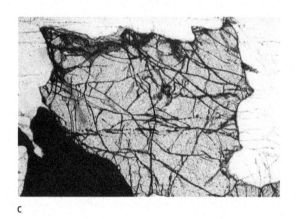

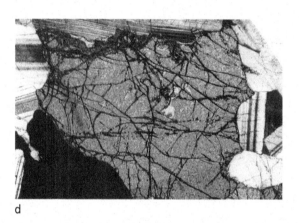

c

d

Photographs c (PP, 40x) and d (XP, 40x) of partially altered Olivine in an olivine gabbro from Wichita Mountain, Oklahoma. The brown alteration product seen in PP light is Iddingsite. The minerals surrounding the Olivine include Plagioclase and a large Opaque grain.

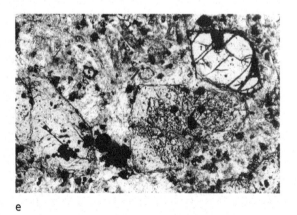

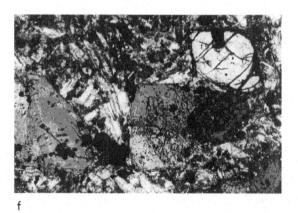

e

f

Photographs e (PP, 40x) and f (XP, 40x) of Olivine and Augite grains in an olivine porphyry basalt from Valmont, Colorado. A relatively unaltered Olivine grain is present in the upper-right corner of the photographs. Below the Olivine grain and to the left are Augite grains. The Augite on the left edge shows a pair of twins. The matrix consists of Plagioclase

PLATE 20: PYROXENES (CONTINUED)

Augite, opaques, Diopside, Calcite, Jadeite, Glaucophane, Riebeckite

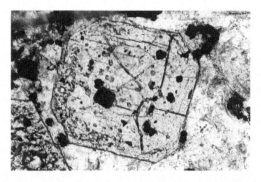

a b

Photographs a (PP, 40x) and b (XP, 40x) of an <u>Augite</u> grain in an olivine porphyry basalt from Valmont, Colorado. This grain yields an optic axis figure, and contains a simple twin.

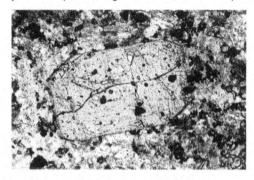
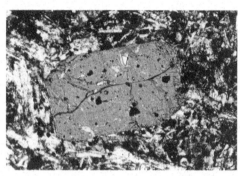

c d

Photographs c (PP, 40x) and d (XP, 40x) of another grain of <u>Augite</u> in the same thin section as photographs a and b. The well-developed prismatic form can be clearly seen.

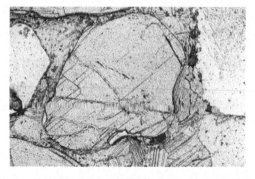
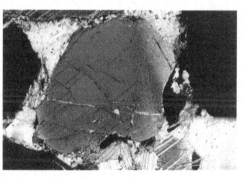

e f

Photographs e (PP, 40x) and f (XP, 40x) of Diopside in a marble from the Adirondack Mountains, New York. The <u>Diopside</u> grain is surrounded by twinned <u>Calcite</u> (high interference colors) and several holes in the thin section (low relief clear areas in PP and black in XP light). Holes form when grains are plucked from thin sections during cutting and polishing, especially near the edges of thin sections. As seen in the PP photograph, the <u>Calcite</u> has been stained light-pink to distinguish it from <u>Dolomite</u>.

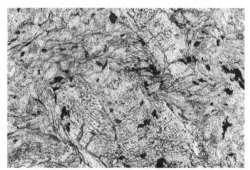

g h

Photographs g (PP, 100x) and h (XP, 100x) of Jadeite and Glaucophane in a blueschist from Panoche Pass, California. In PP light, the <u>Jadeite</u> appears very light green. The thin section also contains <u>Opaques</u> and <u>Amphiboles</u>, including clear to violet <u>Glaucophane</u> and ink-blue <u>Riebeckite</u> (seen in PP light).

125

PLATE 21: PYROXENES (CONTINUED) AND PYROXENOIDS

Pigeonite, Sericite, Hypersthene, Quartz, Biotite, Plagioclase, Wollastonite, Garnet, Diopside

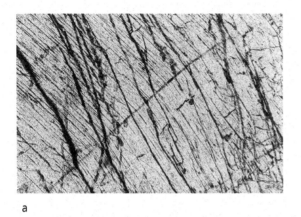
a

b

Photographs a (40x, PP) and b (40x, XP) of Pigeonite in an anorthosite from near Elizabethtown, New York. The <u>Pigeonite</u> has both simple and polysynthetic twinning. <u>Sericite</u> veins cross the <u>Pigeonite</u>.

c

d

Photographs c (PP, 40x) and d (XP, 40x) of Hypersthene in a granulite from near Otter Lake, Quebec. Photograph c shows the green and pink pleochroism, which changes from one color to the other as the stage rotates. Photograph d shows the mineral in XP light along with surrounding grains of <u>Quartz</u>, <u>Biotite</u>, and <u>Plagioclase</u>. Most <u>Hypersthene</u> has less pronounced pleochroism than in this sample.

e

f

Photographs e (PP, 40x) and f (XP, 40x) of Wollastonite in a garnet-wollastonite skarn from Willsboro, Essex County, New York. Two cleavages intersecting at 84° are visible in both photographs. The left sides of the photographs contain <u>Garnet</u> and smaller grains of <u>Diopside</u>.

126

PLATE 22: AMPHIBOLES

Hornblende, Clinopyroxene, Plagioclase, Biotite, Sericite, Glaucophane, Riebeckite, Titanite, Chlorite

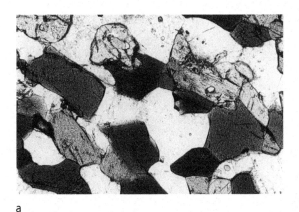

a

b

Photographs a (PP, 100x) and b (XP, 100x) of Hornblende partially altered to Clinopyroxene, perhaps Augite, in an amphibolite from near Cliff Lake along Highway 105 between Vermillion Bay and Perrault Falls, western Ontario. In PP light, the Hornblende is bluish-green to yellow, whereas the Clinopyroxene is pale green. The Hornblende and Clinopyroxene are surrounded by Plagioclase.

c

d

Photographs c (PP, 40x) and d (XP, 40x) of Hornblende in a hornblende syenite from Cuttingsville, Vermont. The photographs show a smaller Hornblende grain penetrating into a larger Hornblende. In PP light, the Hornblende is greenish-brown to yellow. Inclusions in the larger Hornblende include brown Biotite. The 56° and 124° cleavage angles are visible on the right side of the larger grain in PP light. The Hornblendes are surrounded by Plagioclase, which has been partially altered to Sericite.

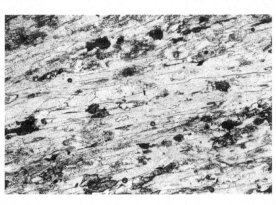

e

f

Photographs e (PP, 40x) and f (XP, 40x) of Glaucophane in a blueschist from Sonoma County, California. In PP light, clear to violet Glaucophane and ink-blue Riebeckite are visible. There are also minor amounts of (high-relief) Titanite (sphene), green Hornblende, and green Chlorite.

PLATE 23: AMPHIBOLES (CONTINUED)

Actinolite, Chlorite, Tremolite, Quartz

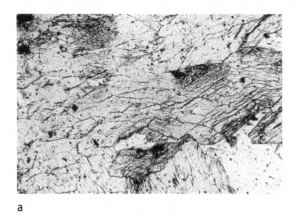

a

b

Photographs a (PP, 40x) and b (XP, 40x) of actinolite in an Actinolite schist from Chester, Vermont. The <u>Actinolite</u> is pale green in PP light. The 56° and 124° cleavage angles are visible in both photographs. <u>Chlorite</u>, showing anomalous blue interference colors, is present in the lower-right-hand corner of the photographs.

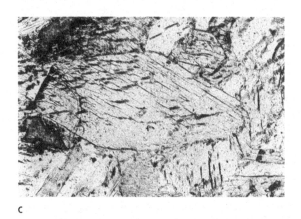

c

d

Photographs c (PP, 40x) and d (XP, 40x) of Tremolite in a marble from the Adirondack Mountains, New York. The <u>Tremolite</u> is clear and colorless in PP light. The 56° and 124° cleavage angles are clearly visible, especially in PP light.

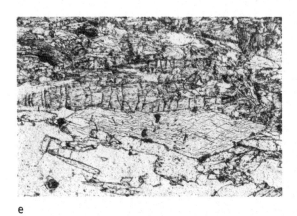

e

f

Photographs e (PP, 100x) and f (XP, 100x) of Tremolite in a marble from the Adirondack Mountains, New York. The <u>Tremolite</u> occurs as needles, rather than as the large blocky grains (photographs c and d). In PP light, the <u>Tremolite</u> is clear and colorless. The 56° and 124° cleavage angles are visible in PP light. <u>Quartz</u> grains are present in the lower-left of the photographs.

PLATE 24: AMPHIBOLES (CONTINUED) AND SHEET SILICATES

Cummingtonite, Garnet, Biotite, Grunerite, Magnetite, Chlorite, Hematite, Talc, Tremolite

a

b

Photographs a (100x, PP) and b (100x, XP) of Cummingtonite needles in a cummingtonite schist from Rockford, South Dakota. The thin section also contains <u>Garnet</u> and small amounts of brown to green <u>Biotite</u>.

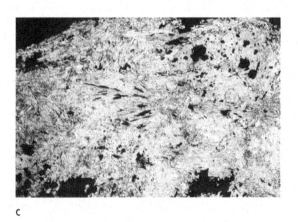

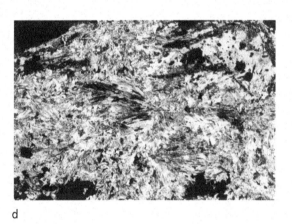

c

d

Photographs c (40x, PP) and d (40x, XP) of Grunerite and opaques in a grunerite-magnetite rock from Michigamme, Michigan. The <u>Grunerite</u> forms a large splay of needles near the center of both photographs. Small amounts of green <u>Chlorite</u> and red <u>Hematite</u> are visible in the upper right-hand corner of the photographs.

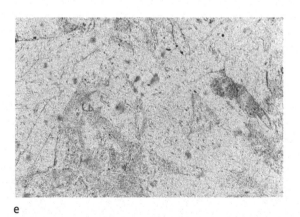

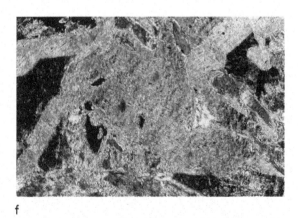

e

f

Photographs e (PP, 100x) and f (XP, 100x) of Talc in a talc-tremolite schist from St. Lawrence County, New York. In XP light, the <u>Talc</u> appears pink and green. Like many sheet silicates, the <u>Talc</u> has "bird's eye extinction"; that is, at extinction the grain does not go solidly black but remains speckled (notice the grain on the left side of photograph f). Small <u>Tremolite</u> grains are also visible at the top center of the photograph in XP light.

Antigorite, Magnetite, Magnesite, Muscovite, Plagioclase, Microcline, Quartz, Biotite

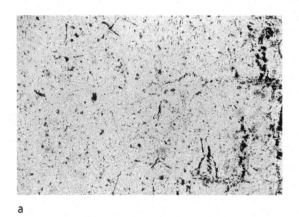
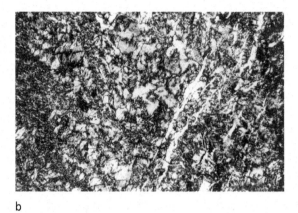

a

b

Photographs a (PP, 40x) and b (XP, 40x) of serpentine in a serpentinite from Rochester, Vermont. The rock largely consists of Serpentine (Antigorite). A few opaques (Magnetite) are visible on the right side of the photograph in PP light. Pearl white veins of Magnesite are visible on the right side of the photograph in XP light.

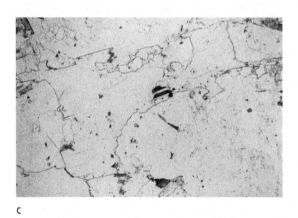

c

d

Photographs c (PP, 40x) and d (XP, 40x) of Muscovite in a biotite granite from Barre, Vermont. In PP light, the Muscovite is clear and colorless. As seen in XP light, the Muscovite is surrounded by Plagioclase at the top of the photograph. A small grain of Microcline with Scotch plaid twinning is seen on the left side of the Muscovite, and Quartz grains, some very fine-grained, on the right.

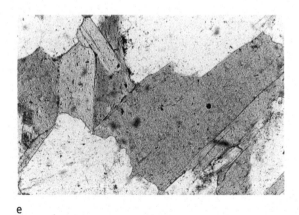

e

f

Photographs e (PP, 100x) and f (XP, 100x) of Biotite showing a bird's eye texture in a diorite from Los Angeles County, California. In PP light, the Biotite flakes are yellowish-brown to brown. The Biotites are surrounded by Plagioclase. Note the lower order interference colors at the thin edges of the large biotite grain.

PLATE 26: MICAS AND OTHER SHEET SILICATES (CONTINUED)

Phlogopite, Calcite, Diopside, Chlorite, Biotite, Plagioclase, Muscovite, Quartz

a

b

Photographs a (PP, 100x) and b (XP, 100x) of Phlogopite surrounded by fine-grained Calcite and Diopside in a marble from Duchess County, New York. In PP light, the <u>Phlogopite</u> is light-brown. In XP light, the <u>Phlogopite</u>, like most <u>Micas</u>, shows "bird's eye extinction."

c

d

Photographs c (PP, 100x) and d (XP, 100x) of Biotite partially altered to Chlorite in a metamorphosed tonalite from San Diego County, California. In PP light, green <u>Chlorite</u> lenses are visible in the brown <u>Biotite</u>. Deep anomalous blue interference colors of <u>Chlorite</u> are seen in XP light. <u>Plagioclase</u> surrounds the sheet silicates.

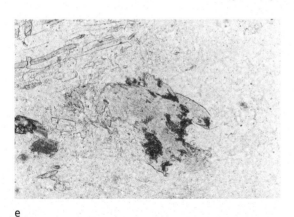

e

f

Photographs e (PP, 100x) and f (XP, 100x) of green Chlorite in a muscovite biotite granite from Concord, New Hampshire. The <u>Chlorite</u> is an alteration product of brown <u>Biotite</u>. A small remnant of the brown <u>Biotite</u> is visible at the top of the large <u>Chlorite</u> grain in the center of photograph e. In XP light, the <u>Chlorite's</u> anomalous blue interference colors are visible. Clear <u>Muscovite</u> grains are present on the left side of the photographs. <u>Plagioclase</u> and <u>Quartz</u> are also present.

PLATE 27: MINERALS COMMON IN METAMORPHIC ROCKS

Epidote, Quartz, Plagioclase, Hematite, Lawsonite, Glaucophane/Riebeckite, opaques

a

b

Photographs a (PP, 40x) and b (XP, 40x) of Epidote from an epidosite, Texas Creek, Colorado. Epidote grains are surrounded by clear Quartz. The larger Epidote grains display compositional zoning in XP light.

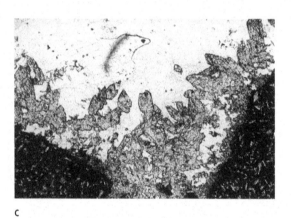

c

d

Photographs c (PP, 40x) and d (XP, 40x) of Epidote in a metamorphosed amygdaloidal basalt from Keweenaw County, Michigan. High relief Epidote grains are pleochroic yellow in PP light and line the rim of a large vesicle in the epidotized basalt. Quartz has filled the rest of the vesicle. The partially altered red basalt, visible in the lower-left and -right corners of the photographs, consists of fine-grained altered Plagioclase and opaque Hematite.

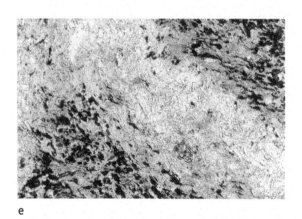

e

f

Photographs e (PP, 100x) and f, (XP, 100x) of a Lawsonite vein in a blueschist from Panoche Pass, California. The small, elongated Lawsonite grains are clear in PP light. Blue Amphibole (Glaucophane or Riebeckite) and Opaques surround the vein.

PLATE 28: MINERALS COMMON IN METAMORPHIC ROCKS (CONTINUED)

Stilbite, opaques, Garnet, Stilpnomelane, Hornblende

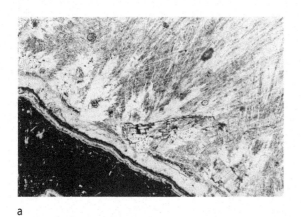

a

b

Photographs a (PP, 40x) and b (XP, 40x) of bladed Stilbite, a Zeolite, in a basalt from Nova Scotia. The <u>Stilbite</u> formed in a vug (a former gas vesicle) in the basalt. An <u>Opaque</u> fine-grained matrix is visible in the lower-left corner of the photographs.

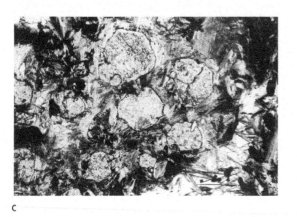

c

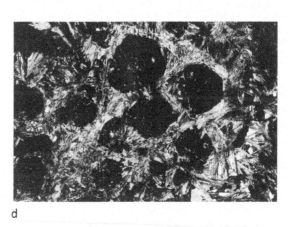

d

Photographs c (PP, 40x) and d (XP, 40x) of Garnet and Stilpnomelane in a stilpnomelane schist from Mendocino County, California. Blades of pleochroic yellowish-brown to brown <u>Stilpnomelane</u> surrounds the pale pink <u>Garnets</u> (PP light).

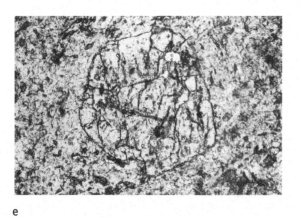

e

f

Photographs e (PP, 40x) and f (XP, 40x) of Garnet in an eclogite from Sonoma County, California. In PP light, the pink <u>Garnet</u> is seen to be surrounded by pale green <u>Omphacite</u> (pyroxene) that has been largely altered to green actinolitic <u>Hornblende</u>.

PLATE 29: ALUMINOUS MINERALS IN METAMORPHIC ROCKS

Andalusite, Sericite, Sillimanite, Muscovite, Quartz, Biotite

a

b

Photographs a (PP, 40x) and b (XP, 40x) of part of a partially altered cross of Andalusite (Chiastolite) in an andalusite schist from Mariposa County, California. The gray <u>Andalusite</u> has been partially altered to <u>Sericite</u> (XP light).

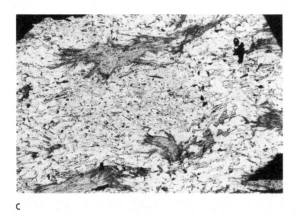

c

d

Photographs c (PP, 40x) and d (XP, 40x) of Andalusite and Sillimanite in a schist from near Savant Lake, western Ontario. In XP light, the gray <u>Andalusite</u> is difficult to see because of <u>Quartz</u> inclusions and surrounding <u>Quartz</u> grains but most of the center of the photo is Andalusite. Both <u>Andalusite</u> and <u>Quartz</u> have similar interference colors. The fine-grained <u>Sillimanite</u> (<u>Fibrolite</u>) appears as brown fibrous bundles in PP light. <u>Muscovite</u> is also visible in the thin section.

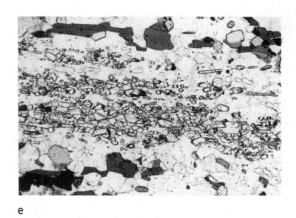

e

f

Photographs e (PP, 40x) and f (XP, 40x) of Sillimanite in a schist from Kazabazua, Quebec. In PP light, the <u>Sillimanite</u> appears as high relief, clear, blocky crystals crossing the center of the photograph. Reddish-brown <u>Biotite</u> and <u>Quartz</u> are also present. In XP light, the <u>Sillimanite</u> grains are mostly at extinction.

PLATE 30: ALUMINOUS MINERALS IN METAMORPHIC ROCKS (CONTINUED)

Kyanite, Plagioclase, Sericite, Staurolite, Quartz, Chloritoid, Garnet, Graphite, Muscovite

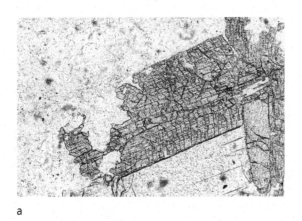
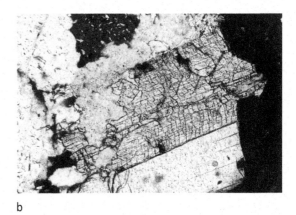

a

b

Photographs a (PP, 100x) and b (XP, 100x) of Kyanite in a kyanite quartzite from Ogilby, California. The highly fractured <u>Kyanite</u> is clear and has high relief. Some of the <u>Kyanite</u> along the right edge of the photograph has been plucked, leaving holes in the thin section. The remaining <u>Kyanite</u> along the right edge is at extinction. The <u>Kyanite</u> is surrounded by <u>Quartz</u> and partially sericitized <u>Plagioclase</u>.

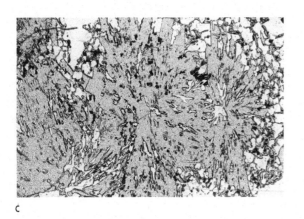

c

d

Photographs c (PP, 40x) and d (XP, 40x) of Staurolite in a staurolite quartzite from Petaca, New Mexico. In PP light, the <u>Staurolite</u> is light yellow and pleochroic. <u>Quartz</u> is present as inclusions and as surrounding grains.

e

f

Photographs e (PP, 40x) and f (XP, 40x) of Chloritoid in a schist from Duchess County, New York. An elongated <u>Chloritoid</u> grain crosses the center of the photographs and is surrounded by fine-grained <u>Muscovite</u>, <u>Quartz</u>, and one large <u>Garnet</u>. Long thin opaques, likely <u>Graphite</u>, are also present.

PLATE 31: ALUMINOUS MINERALS IN METAMORPHIC ROCKS (CONTINUED)

Spinel, Magnetite, Biotite, Quartz, Plagioclase, Cordierite, Sillimanite, Garnet

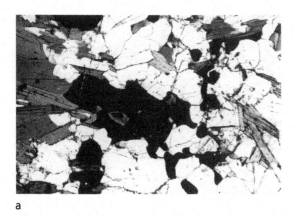

a

b

Photographs a (PP, 40x) and b (XP, 40x) of Spinel contained in Magnetite from a gneiss collected near Manitou Falls, western Ontario. Green Spinel (Pleonaste) is visible within the Magnetite in PP light. Under high-temperature conditions, the Spinel and Magnetite may have existed as a solid solution and then separated into the distinct minerals during cooling. The Spinel and Magnetite are surrounded by brown (in PP light) Biotite, Quartz, and Plagioclase.

c

d

Photographs c (PP, 40x) and d (XP, 40x) of Spinel within Cordierite in a gneiss from near Manitou Falls, western Ontario. Besides the green Spinel (Pleonaste), fibrous Sillimanite is present as inclusions in the Cordierite. Minor red and yellow Biotites are also present in the thin section. In XP light, twins are clearly visible in the Cordierite.

e

f

Photographs e (PP, 40x) and f (XP, 40x) of Sillimanite and Cordierite in a gneiss from near Manitou Falls, western Ontario. In PP light, the Sillimanite appears as needles in a heart-shaped group within the Cordierite. In XP light, Cordierite twins are clearly visible. The thin section also contains abundant yellowish- to reddish-brown (in PP light) Biotite, Quartz, Plagioclase, and a Garnet in the lower-right of the photographs.

Cordierite, Zircon, Biotite, Hypersthene, Corundum, Sapphirine

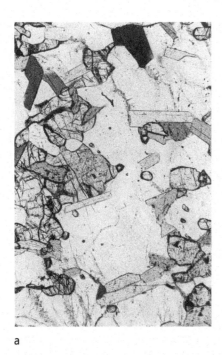

a   b

Photographs a (40x, PP) and b (40x, XP) of Cordierite with Zircon inclusions in a gneiss from near Kazabazua, Quebec. In PP light, yellow halos are often seen around the <u>Zircons</u>. The <u>Cordierites</u> show visible twinning in XP light. The thin section also contains brown (in PP light) <u>Biotite,</u> and <u>Hypersthene</u>.

c   d

Photographs c (PP, 40x) and d (XP, 40x) of Corundum, Sapphirine, and Hypersthene in an aluminous gneiss from Natawahunan Lake, Manitoba. In PP light, a very large hexagonal <u>Corundum</u> grain can be seen. The thin section is cut nearly perpendicular to the optic axis of the hexagonal <u>Corundum</u> grain. Therefore, the grain appears at extinction in XP light. Blue <u>Sapphirine</u> and pink <u>Hypersthene</u>, along with minor brown <u>Biotite</u>, are also present in the thin section.

Scapolite, Apatite, Clinopyroxene, Titanite, Vesuvianite, Calcite, Pyrite, Tremolite

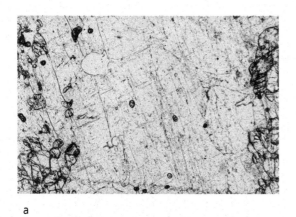

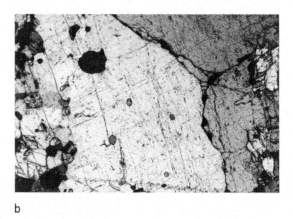

a

b

Photographs a (PP, 40x) and b (XP, 40x) of Scapolite in a metabasite from Fort Coulonge, Quebec. In XP light, the Scapolite has low (gray) interference colors and near 90° cleavage. The Scapolite contains Apatite inclusions. Clinopyroxenes and a few small grains of Titanite (sphene) appear on the extreme left and right edges of the photographs.

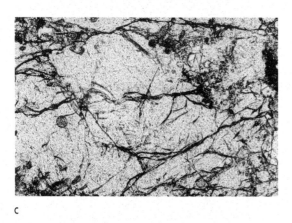

c

d

Photographs c (PP, 40x) and d (XP, 40x) of Vesuvianite (Idocrase) and Calcite in a skarn from Lake Chaco, Chihuahua, Mexico. The Vesuvianite shows anomalous blue interference colors (Photograph d). Calcite grains are present in the upper-left corner and on the right side of the photographs. Fractures in the center of the photographs contain smaller Calcite grains.

e

f

Photographs e (PP, 40x) and f (reflected light, 40x) of Pyrite in a marble from the Adirondack Mountains, New York. In PP light, the Pyrite is opaque (black) and shows no details. In reflected light (f), the gold color is visible. Surrounding the Pyrite is red-stained Calcite, possible elongated white grains of Tremolite, and a small hole in the thin section with light-green epoxy. The red stain distinguishes Calcite from other carbonates.

PLATE 34: OPAQUE MINERALS IN THIN SECTION (CONTINUED)

Chalcopyrite, Hornblende, Quartz, Plagioclase, Biotite, Magnetite, Ilmenite, Hematite

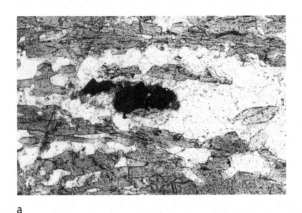

a

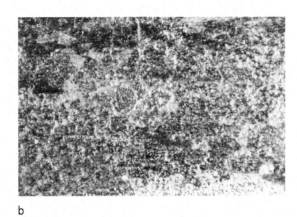

b

Photographs a (PP, 100x) and b (reflected light, 100x) of Chalcopyrite in an amphibolite from near Osnaburgh House, Ontario. The yellowish-gold color of the Chalcopyrite is visible in photograph b. In PP light (photograph a), the opaque Chalcopyrite grain is seen to be surrounded by olive green to bluish-green Hornblende, Plagioclase, Quartz, and small amounts of brown Biotite.

c

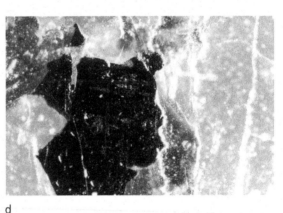

d

Photographs c (PP, 200x) and d (reflected light, 200x) of an opaque grain consisting of a mixture of Magnetite and Ilmenite in an amphibolite from Otter Lake, Quebec. The Magnetite-Ilmenite grain is surrounded by brown Biotite and clear Quartz, as seen in PP light. Usually opaques are best distinguished in reflected light. However, in this case, the exsolution of Ilmenite from Magnetite is more clearly seen in dim PP light (photograph c).

e

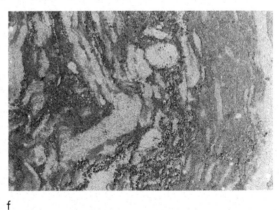

f

Photographs e (PP, 40x) and f (reflected light, 40x) of banded iron formation from Negaunee, Michigan. In PP light, clear Quartz is visible, but the opaques cannot be distinguished from each other. In reflected light, red Hematite is easily distinguished from the bluish specular variety of Hematite. Quartz is whitish to light-brown in the reflected light.

APPENDIX A

Common Opaque Minerals

Mineral	Crystal System
chalcopyrite	tetragonal
chromite	cubic
graphite	hexagonal
ilmenite	hexagonal
magnetite	cubic
pyrite	cubic
pyrrhotite	hexagonal

Isotropic Minerals Ordered by Refractive Index

(Minerals in **bold** are described in detail in this book; the others are rare or difficult to distinguish in thin sections.)

Mineral	Crystal System	Mean Refractive Index
fluorite	cubic	1.434
sodalite	cubic	1.480
analcime	cubic	1.490
sylvite	cubic	1.490
halite	cubic	1.540
garnet	cubic	1.710–1.870
spinel	cubic	1.720–1.740
periclase	cubic	1.730–1.736
sphalerite	cubic	2.390–2.420
diamond	cubic	2.419

Uniaxial Minerals Sorted by Optic Sign and Ordered by Refractive Index

(Minerals in **bold** are described in detail in this book; the others are rare or difficult to distinguish in thin section.)

►UNIAXIAL (−) MINERALS

Mineral	Crystal System	Mean Refractive Index (R.I.)	Mean Birefringence
chabazite	hexagonal	1.48	0.002–0.005
calcite	hexagonal	1.48–1.65	0.172
dolomite	hexagonal	1.50–1.67	0.177–0.185
magnesite	hexagonal	1.50–1.70	0.191
nepheline	hexagonal	1.54	0.003–0.005
scapolite	tetragonal	1.54–1.59	0.002–0.039
ankerite	hexagonal	1.54–1.75	0.202
beryl	hexagonal	1.56	0.004–0.008
rhodochrosite	hexagonal	1.59–1.81	0.219
tourmaline	hexagonal	1.62–1.67	0.021–0.029
apatite	hexagonal	1.63	0.003
siderite	hexagonal	1.63–1.87	0.239–0.242
vesuvianite	tetragonal	1.70	0.004–0.006
corundum	hexagonal	1.76	0.008–0.009
hematite	hexagonal	3.00–3.20	0.210–0.280

UNIAXIAL (+) MINERALS

Mineral	Crystal System	Mean Refractive Index (R.I.)	Mean Birefringence
leucite	tetragonal	1.51	0.001
apophyllite	tetragonal	1.53	0.002
quartz	hexagonal	1.54	0.009
brucite	hexagonal	1.57–1.58	0.015–0.021
zircon	tetragonal	1.93–1.99	0.060–0.062
rutile	tetragonal	2.61–2.90	0.286–0.287
cassiterite	tetragonal	2.00–2.10	0.090

APPENDIX D

Biaxial Minerals Sorted by Optic Sign and Ordered by Refractive Index

(Minerals in **bold** are described in detail in this book; the others are rare or difficult to distinguish in thin sections.)

▶BIAXIAL (−) MINERALS

Minerals	Crystal System	2V°	Mean Refractive Index (R.I.)	Mean Birefringence
borax	monoclinic	40	1.44–1.47	0.025
kernite	monoclinic	80	1.45–1.48	0.034
montmorillomite	monoclinic	0–30	1.48–1.52	0.025–0.031
stilbite	monoclinic	30–49	1.50	0.010
chrysotile	monoclinic	0–50	1.50–1.52	0.004–0.014
chalcanthite	triclinic	56	1.51–1.54	0.029
orthoclase	monoclinic	60–65	1.52	0.005–0.007
sanidine	monoclinic	variable	1.52	0.005–0.007
microcline	triclinic	77–84	1.52	0.006
strontianite	orthorhombic	7	1.52–1.66	0.148
witherite	orthorhombic	16	1.52–1.67	0.148
lepidolite	monoclinic	0–60	1.53–1.55	0.020–0.040
plagioclase	triclinic	75–90	1.53–1.59	0.007–0.013
aragonite	orthorhombic	18	1.53–1.68	0.155
cordierite	orthorhombic	62–90	1.54–1.56	0.008–0.018
talc	monoclinic	0–30	1.54–1.58	0.046–0.050
stilpnomelane	monoclinic	≅0	0 1.54–1.75	0.030–0.110
pyrophyllite	triclinic	53–62	1.55–1.60	0.045–0.048
kaolinite	triclinic	24–50	1.56	0.007
muscovite	monoclinic	35–50	1.56–1.60	0.036–0.054
antigorite	monoclinic	27–60	1.57	0.006–0.009
biotite	monoclinic	0–33	1.57–1.61	0.028–0.081
anthophyllite	orthorhombic	65–90	1.60–1.63	0.017–0.026
tremolite-actinolite	monoclinic	74–85	1.61–1.63	0.016–0.030
lazulite	monoclinic	70	1.61–1.64	0.031

continues

►BIAXIAL (−) MINERALS (Continued)

Minerals	Crystal System	2V°	Mean Refractive Index (R.I.)	Mean Birefringence
wollastonite	triclinic	36–42	1.63	0.014
andalusite	orthorhombic	83–85	1.63	0.009–0.011
margarite	monoclinic	45	1.63–1.65	0.013
monticellite	orthorhombic	72–82	1.64–1.66	0.020
hornblende	monoclinic	52–85	1.65–1.67	0.020
glancophane-riebeckite	monoclinic	10–90	1.65–1.67	0.006–0.029
malachite	monoclinic	43	1.65–1.90	0.254
clinozoisite	monoclinic	14–90	1.67–1.73	0.005–0.015
kaersutite	monoclinic	68–82	1.68–1.73	0.020–0.080
grunerite	monoclinic	84–90	1.68–1.73	0.043
kyanite	triclinic	82–83	1.72	0.014–0.017
epidote	monoclinic	64–89	1.71–1.80	0.015–0.048
cerussite	orthorhombic	9	1.80–2.07	0.274
lepidocrocite	orthorhombic	83	1.94–2.51	0.570
goethite	orthorhombic	0–27	2.26–2.52	0.150
realgar	monoclinic	41	2.50–2.80	0.166

►BIAXIAL (+) MINERALS

Minerals	Crystal System	2V°	Mean Refractive Index (R.I.)	Mean Birefringence
natrolite	orthorhombic	58–64	1.48	0.012
tridymite	orthorhombic	70	1.48	0.003
montmorillomite	monoclinic	0–30	1.48–1.52	0.025–0.031
heulandite	monoclinic	0–55	1.50	0.005–0.006
chrysotile	monoclinic	0–50	1.50–1.52	0.004–0.014
gypsum	monoclinic	58	1.52	0.010
plagioclase	triclinic	75–90	1.53–1.59	0.007–0.013
cordierite	orthorhombic	65–90	1.54–1.56	0.008–0.018
gibbsite	monoclinic	0–40	1.57–1.59	0.020
anhydrite	orthorhombic	42	1.57–1.61	0.044
colemanite	monoclinic	56	1.58–1.61	0.028
coesite	monoclinic	64	1.59	0.010
pectolite	triclinic	35–63	1.59–1.63	0.040
topaz	orthorhombic	44–66	1.61	0.008–0.010
turquoise	triclinic	40	1.61–1.65	0.040
celestite	orthorhombic	51	1.62	0.009
prehnite	orthorhombic	65–69	1.62–1.65	0.021–0.039
barite	orthorhombic	26–38	1.64	0.012
cummingtonite	monoclinic	80–90	1.64–1.67	0.030–0.043
boracite	orthorhombic	82	1.64–1.67	0.011
jadeite	monoclinic	68–72	1.65–1.67	0.012–0.023
spodumene	monoclinic	60–80	1.65–1.67	0.020
sillimanite	orthorhombic	21–30	1.65–1.68	0.021
hypersthene	orthorhombic	54	1.66	0.008
lawsonite	orthorhombic	79–85	1.66–1.68	0.020
diopside	monoclinic	56–63	1.66–1.75	0.031
diaspore	orthorhombic	85	1.68–1.75	0.040
pigeonite	monoclinic	38–44	1.69–1.72	0.025
chloritoid	monoclinic	36–63	1.71–1.72	0.010–0.012
rhodonite	triclinic	63–76	1.71–1.73	0.013
antlerite	orthorhombic	53	1.72–1.78	0.063
azurite	monoclinic	68	1.73–1.83	0.106
staurolite	monoclinic	79–90	1.75	0.013–0.015

Minerals	Crystal System	2V°	Mean Refractive Index (R.I.)	Mean Birefringence
chrysoberyl	orthorhombic	0–90	1.74	0.010
monazite	monoclinic	10–20	1.78–1.85	0.050
titanite (sphene)	monoclinic	23–50	1.86–2.10	0.108–0.160
anglesite	orthorhombic	75	1.87–1.89	0.017
sulfur	orthorhombic	69	1.95–2.24	0.287
wolframite	monoclinic	73–79	2.10–2.50	0.130–0.150
manganite	monoclinic	≅0	2.20–2.50	0.290
orpiment	monoclinic	76	2.40–3.00	0.620

Minerals Ordered by Interference Colors and Sorted by Optic System and Optic Sign

(Minerals in **bold** are described in detail in this book; the others are rare or difficult to distinguish in thin sections.)

▶MINERALS WITH VERY LOW BIREFRINGENCE
(Normally First-Order Gray, White Interference Colors)

Mineral	Crystal System	Optical System and Optic Sign	2V°	Mean Refractive Index (R.I.)	Mean Birefringence
apatite	hexagonal	uniaxial −		1.63	0.003
beryl	hexagonal			1.56	0.002–0.005
chabazite	hexagonal			1.48	0.007
corundum	hexagonal			1.76	0.005–0.009
nepheline	hexagonal			1.54	0.002–0.039
scapolite	tetragonal			1.54–1.59	0.002–0.004
vesuvianite	tetragonal			1.70	0.002
apophyllite	tetragonal	uniaxial +		1.53	0.002
leucite	tetragonal			1.51	0.001
quartz	hexagonal			1.54	0.009
antigorite	monoclinic	biaxial −	27–60	1.57	0.006–0.009
clinozoisite	monoclinic		14–90	1.67–1.73	0.005–0.015
kaolinite	triclinic		24–50	1.56	0.005
microcline	triclinic		77–84	1.52	0.006
orthoclase	monoclinic		60–65	1.52	0.005–0.007
sanidine	monoclinic		variable	1.52	0.005–0.007
chlorite	monoclinic	biaxial + or −	0–50	1.56–1.61	0.006–0.020
heulandite	monoclinic	biaxial +	0–55	1.50	0.005
tridymite	orthorhombic		70	1.48	0.003

▶MINERALS WITH LOW BIREFRINGENCE
(Up To First-Order Yellow or Red Interference Colors)

Mineral	Crystal System	Optical System and Optic Sign	2V°	Mean Refractive Index (R.I.)	Mean Birefringence
corundum	hexagonal	uniaxial −		1.76	0.008–0.009
andalusite	orthorhombic	biaxial −	83–85	1.63	0.009–0.011
glaucophane–riebeckite	monoclinic		10–90	1.65–1.67	0.006–0.029
kyanite	triclinic		82–83	1.72	0.014–0.017
margarite	monoclinic		45	1.63–1.65	0.013
stilbite	monoclinic		30–49	1.50	0.010
wollastonite	triclinic		36–42	1.63	0.014
cordierite	orthorhombic	biaxial + or −	65–90	1.54–1.56	0.008–0.018
orthopyroxene	orthorhombic		75–90	1.65–1.72	0.008–0.016
plagioclase	triclinic		63–76	1.53–1.59	0.007–0.013
barite	orthorhombic	biaxial +	26–38	1.64	0.012
boracite	orthorhombic		82	1.64–1.67	0.011
celestite	orthorhombic		51	1.62	0.009
chloritoid	monoclinic		36–63	1.72	0.010–0.012
chrysoberyl	orthorhombic		0–90	1.74–1.75	0.010
chrysotile	monoclinic		0–50	1.50–1.52	0.004–0.014
coesite	monoclinic		64	1.59	0.010
gypsum	monoclinic		42–90	1.52	0.010
hypersthene	orthorhombic		54	1.66	0.008
natrolite	orthorhombic		58–64	1.48	0.012
rhodonite	triclinic		58–90	1.71–1.73	0.013
staurolite	monoclinic		79–90	1.75	0.013–0.015
topaz	orthorhombic		44–66	1.61	0.008–0.010

▶MINERALS WITH MODERATE TO HIGH BIREFRINGENCE
(Second to Fourth Order Interference Colors)

Mineral	Crystal System	Optical System and Optic Sign	2V°	Mean Refractive Index (R.I.)	Mean Birefringence
tourmaline	hexagonal	uniaxial −		1.62–1.67	0.021–0.029
brucite	hexagonal	uniaxial +		1.57–1.58	0.015–0.021
anthophyllite	orthorhombic	biaxial −	65–90	1.60–1.63	0.017–0.026
biotite	monoclinic		0–33	1.57–1.61	0.028–0.081
borax	monoclinic		40	1.44–1.47	0.025
chalcanthite	triclinic		56	1.51–1.54	0.029
epidote	monoclinic		64–89	1.71–1.80	0.015–0.048
grunerite	monoclinic		84–90	1.68–1.73	0.043
hornblende	monoclinic		52–85	1.65–1.67	0.020
kaersutite	monoclinic		68–82	1.68–1.73	0.020–0.080
kernite	monoclinic		80	1.45–1.48	0.034
lazulite	monoclinic		70	1.61–1.64	0.031
lepidolite	monoclinic		0–60	1.53–1.55	0.020–0.040
monticellite	orthorhombic		72–82	1.64–1.66	0.020
muscovite	monoclinic		35–50	1.56–1.60	0.036–0.054
pyrophyllite	triclinic		53–62	1.55–1.60	0.045–0.048
talc	monoclinic		0–30	1.54–1.58	0.046–0.050
tremolite–actinolite	monoclinic		74–85	1.61–1.63	0.016–0.030
montmorillonite	monoclinic	biaxial + or −	0–30	1.48–1.52	0.025–0.031
olivine	orthorhombic		47–90	1.63–1.88	0.035–0.051
anglesite	orthorhombic	biaxial +	75	1.87–1.89	0.017
anhydrite	orthorhombic		42	1.57–1.61	0.044
augite	monoclinic		25–60	1.68–1.75	0.024–0.029
colemanite	monoclinic		56	1.58–1.61	0.028
cummingtonite	monoclinic		80–90	1.64–1.67	0.030–0.043
diaspore	orthorhombic		85	1.68–1.75	0.040
diopside	monoclinic		56–63	1.66–1.75	0.031
gibbsite	monoclinic		0–40	1.57–1.59	0.020
jadeite	monoclinic		68–72	1.65–1.67	0.012–0.023
lawsonite	orthorhombic		79–85	1.66–1.68	0.020
monazite	monoclinic		10–20	1.78–1.85	0.050
pectolite	triclinic		35–63	1.59–1.63	0.040
pigeonite	monoclinic		38–44	1.69–1.72	0.025
prehnite	orthorhombic		65–69	1.62–1.65	0.021–0.039
sillimanite	orthorhombic		21–30	1.65–1.68	0.021
spodumene	monoclinic		60–80	1.65–1.67	0.020
turquoise	triclinic		40	1.61–1.65	0.040

▶MINERALS WITH EXTREME BIREFRINGENCE
(Fourth-Order and Above Interference Colors)

Mineral	Crystal System	Optical System and Optic Sign	2V°	Mean Refractive Index (R.I.)	Mean Birefringence
ankerite	hexagonal	uniaxial −		1.54–1.75	0.202
calcite	hexagonal			1.48–1.65	0.172
dolomite	hexagonal			1.50–1.67	0.177–0.185
hematite	hexagonal			3.00–3.20	0.210–0.280
magnesite	hexagonal			1.50–1.70	0.191
rhodochrosite	hexagonal			1.59–1.81	0.219
siderite	hexagonal			1.63–1.87	0.239–0.242
cassiterite	tetragonal	uniaxial +		2.00–2.10	0.090
rutile	tetragonal			2.61–2.90	0.286–0.287
zircon	tetragonal			1.93–1.99	0.060–0.062
aragonite	orthorhombic	biaxial −	18	1.53–1.68	0.155
cerussite	orthorhombic		9	1.80–2.07	0.274
goethite	orthorhombic		0–27	2.26–2.52	0.150
lepidocrocite	orthorhombic		83	1.94–2.51	0.570
malachite	monoclinic		43	1.65–1.90	0.254
realgar	monoclinic		41	2.50–2.80	0.166
strontianite	orthorhombic		7	1.52–1.66	0.148
witherite	orthorhombic		16	1.52–1.67	0.148
antlerite	orthorhombic	biaxial +	53	1.72–1.78	0.063
azurite	monoclinic		68	1.73–1.83	0.106
manganite	monoclinic		≅0	2.20–2.50	0.290
orpiment	monoclinic		76	2.40–3.00	0.620
sulfur	orthorhombic		69	1.95–2.24	0.287
titanite (sphene)	monoclinic		23–50	1.86–2.10	0.108–0.160
wolframite	monoclinic		73–79	2.10–2.50	0.130–0.150

Alphabetical List of Minerals and Mineral Properties

Mineral name	Crystal System	Optical System	Optic Sign	2V°	Mean Refractive Index (R. I.)	Mean Birefringence
actimolite (see tremolite-actinolite)						
analcime	cubic	isotropic			1.49	
andalusite	orthorhombic	biaxial	−	83–85	1.63	0.009–0.011
anglesite	orthorhombic	biaxial	+	75	1.87–1.89	0.017
anhydrite	orthorhombic	biaxial	+	42	1.57–1.61	0.044
ankerite	hexagonal	uniaxial	−		1.54–1.75	0.202
anorthoclase (see sanidine)						
anthophyllite	orthorhombic	biaxial	−	65–90	1.60–1.63	0.017–0.026
antigorite	monoclinic	biaxial	−	27–60	1.57	0.006–0.009
antlerite	orthorhombic	biaxial	+	53	1.72–1.78	0.063
apatite	hexagonal	uniaxial	−		1.63	0.003
apophyllite	tetragonal	uniaxial	+		1.53	0.002
aragonite	orthorhombic	biaxial	−	18	1.53–1.68	0.155
augite	monoclinic	biaxial	+	25–60	1.68–1.75	0.024–0.029
azurite	monoclinic	biaxial	+	68	1.73–1.83	0.106
barite	orthorhombic	biaxial	+	26–38	1.64	0.012
beryl	hexagonal	uniaxial	−		1.56	0.004–0.008
biotite	monoclinic	biaxial	−	0–33	1.57–1.61	0.028–0.081
boracite	orthorhombic	biaxial	+	82	1.64–1.67	0.011
borax	monoclinic	biaxial	−	40	1.44–1.47	0.025
brucite	hexagonal	uniaxial	+		1.57–1.58	0.015–0.021
calcite	hexagonal	uniaxial	−		1.48–1.65	0.172
cassiterite	tetragonal	uniaxial	+		2.00–2.10	0.090
celestite	orthorhombic	biaxial	+	51	1.62	0.009
cerussite	orthorhombic	biaxial	−	9	1.80–2.07	0.274
chabazite	hexagonal	uniaxial	−		1.48	0.002–0.005
chalcanthite	triclinic	biaxial	−	56	1.51–1.54	0.029
chalcedony (see quartz)						
chalcopyrite	tetragonal	opaque				

continues

Mineral name	Crystal System	Optical System	Optic Sign	2V°	Mean Refractive Index (R. I.)	Mean Birefringence
chlorite	monoclinic	biaxial	+/−	0–50	1.56–1.61	0.006–0.020
chloritoid	monoclinic	biaxial	+	36–63	1.72	0.010–0.012
chromite	cubic	opaque				
chrysoberyl	orthorhombic	biaxial	+	0–90	1.74–1.75	0.010
chrysotile	monoclinic	biaxial	+	0–50	1.50–1.52	0.004–0.014
clinozoisite	monoclinic	biaxial	−	14–90	1.67–1.73	0.005–0.015
coesite	monoclinic	biaxial	+	64	1.59	0.010
colemanite	monoclinic	biaxial	+	56	1.58–1.61	0.028
cordierite	orthorhombic	biaxial	+/−	65–90	1.54–1.56	0.008–0.018
corundum	hexagonal	uniaxial	−		1.76	0.005–0.009
crossite (see glaucophane-riebeckite)						
cummingtonite	monoclinic	biaxial	+	80–90	1.64–1.67	0.030–0.043
diamond	cubic	isotropic			2.419	
diaspore	orthorhombic	biaxial	+	85	1.68–1.75	0.040
diopside	monoclinic	biaxial	+	56–63	1.66–1.75	0.031
dolomite	hexagonal	uniaxial	−		1.50–1.67	0.177–0.185
enstatite (see orthopyroxene)						
epidote	monoclinic	biaxial	−	64–89	1.71–1.80	0.015–0.048
fluorite	cubic	isotropic			1.434	
garnet	cubic	isotropic			1.71–1.87	
gibbsite	monoclinic	biaxial	+	0–40	1.57–1.59	0.020
glaucophane-riebeckite	monoclinic	biaxial	−	10–90	1.65–1.67	0.006–0.029
geothite	orthorhombic	biaxial	−	0–27	2.26–2.52	0.150
graphite	hexagonal	opaque				
grunerite	monoclinic	biaxial	−	84–90	1.68–1.73	0.043
gypsum	monoclinic	biaxial	+	58	1.52	0.010
halite	cubic	isotropic			1.54	
hematite	hexagonal	uniaxial	−		3.00–3.20	0.210–0.280
heulandite	monoclinic	biaxial	+	0–55	1.50	0.005
hornblende	monoclinic	biaxial	−	52–85	1.65–1.67	0.020
hypersathene (see orthopyroxene)						
ilmenite	hexagonal	opaque				
jadeite	monoclinic	biaxial	+	68–72	1.65–1.67	0.012–0.023
K-feldspar (see orthoclase, monoclinic or sanidine)						
kaersutite	monoclinic	biaxial	−	68–82	1.68–1.73	0.020–0.080
kaolinite	triclinic	biaxial	−	24–50	1.56	0.007
kernite	monoclinic	biaxial	−	80	1.45–1.48	0.034
kyanite	triclinic	biaxial	−	82–83	1.72	0.014–0.017
lawsonite	orthorhombic	biaxial	+	79–85	1.66–1.68	0.020
lazulite	monoclinic	biaxial	−	70	1.61–1.64	0.031
lepidocrocite	orthorhombic	biaxial	−	83	1.94–2.51	0.570
lepidolite	monoclinic	biaxial	−	0–60	1.53–1.55	0.020–0.040
leucite	tetragonal	uniaxial	+		1.51	0.001
magnesite	hexagonal	uniaxial	−		1.50–1.70	0.191
magnetite	cubic	opaque				
malachite	monoclinic	biaxial	−	43	1.65–1.90	0.254
manganite	monoclinic	biaxial	+	≅0	2.20–2.50	0.290
margarite	monoclinic	biaxial	−	45	1.63–1.65	0.013
microcline	triclinic	biaxial	−	77–84	1.52	0.006
monazite	monoclinic	biaxial	+	10–20	1.78–1.85	0.050

Mineral name	Crystal System	Optical System	Optic Sign	2V°	Mean Refractive Index (R. I.)	Mean Birefringence
monticellite	orthorhombic	biaxial	−	72–82	1.64–1.66	0.020
montmorillonite	monoclinic	biaxial	+/−	0–30	1.48–1.52	0.025–0.031
muscovite	monoclinic	biaxial	−	35–50	1.56–1.60	0.036–0.054
natrolite	orthorhombic	biaxial	+	58–64	1.48	0.012
nepheline	hexagonal	uniaxial	−		1.54	0.003–0.005
olivine	orthorhombic	biaxial	+/−	47–90	1.63–1.88	0.035–0.051
orpiment	monoclinic	biaxial	+	76	2.40–3.00	0.620
orthoclase	monoclinic	biaxial	−	60–65	1.52	0.005–0.007
orthopyroxene	orthorhombic	biaxial	+/−	58–90	1.65–1.72	0.008–0.016
pectolite	triclinic	biaxial	+	35–63	1.59–1.63	0.040
periclase	cubic	isotropic			1.73–1.736	
phlogopite (see biotite)						
pigeonite	monoclinic	biaxial	+	38–44	1.69–1.72	0.025
plagioclase	triclinic	biaxial	+/−	75–90	1.53–1.59	0.007–0.013
prehnite	orthorhombic	biaxial	+	65–69	1.62–1.65	0.021–0.039
pyrite	cubic	opaque				
pyrophyllite	triclinic	biaxial	−	53–62	1.55–1.60	0.045–0.048
pyrrhotite	hexagonal	opaque				
quartz	hexagonal	uniaxial	+		1.54	0.009
realgar	monoclinic	biaxial	−	41	2.50–2.80	0.166
rhodochrosite	hexagonal	uniaxial	−		1.59–1.81	0.219
rhodonite	triclinic	biaxial	+	63–76	1.71–1.73	0.013
riebeckite (see glaucophane-riebeckite)						
rutile	tetragonal	uniaxial	+		2.61–2.90	0.286–0.287
sanidine	monoclinic	biaxial	−	variable	1.52	0.005–0.007
scapolite	tetragonal	uniaxial	−		1.54–1.59	0.002–0.004
siderite	hexagonal	uniaxial	−		1.63–1.87	0.239–0.242
sillimanite	orthorhombic	biaxial	+	21–30	1.65–1.68	0.021
sodalite	cubic	isotropic			1.48	
sphalerite	cubic	isotropic			2.39–2.42	
spinel	cubic	isotropic			1.72–1.74	
spodumene	monoclinic	biaxial	+	60–80	1.65–1.67	0.020
staurolite	monoclinic	biaxial	+	79–90	1.75	0.013–0.015
stilbite	monoclinic	biaxial	−	30–49	1.50	0.010
stilpnomelane	monoclinic	biaxial	−	≅0	1.54–1.75	0.030–0.110
strontianite	orthorhombic	biaxial	−	7	1.52–1.66	0.148
sulfur	orthorhombic	biaxial	+	69	1.95–2.24	
sylvite	cubic	isotropic			1.49	
talc	monoclinic	biaxial	−	0–30	1.54–1.58	0.046–0.050
titanite (sphene)	monoclinic	biaxial	+	23–50	1.86–2.10	0.108–0.160
topaz	orthorhombic	biaxial	+	44–66	1.61	0.008–0.010
tourmaline	hexagonal	uniaxial	−		1.62–1.67	0.021–0.029
tremelite-actinolite	monoclinic	biaxial	−	74–85	1.61–1.63	0.016–0.030
tridymite	orthorhombic	biaxial	+	70	1.48	0.003
turquoise	triclinic	biaxial	+	40	1.61–1.65	0.040
vesuvianite	tetragonal	uniaxial	−		1.70	0.004–0.006
witherite	orthorhombic	biaxial	−	16	1.52–1.67	0.148
wolframite	monoclinic	biaxial	+	73–79	2.10–2.50	0.130–0.15
wollastonite	triclinic	biaxial	−	36–42	1.63	0.014
zircon	tetragonal	uniaxial	+		1.93–1.99	0.060–0.062
zoisite (see clinozoisite)						

Color Photographs

Plate	Lithology or Major Mineral	Major Minerals Seen in Photographs	Minor Minerals	Page #
1e–f	scoria	opaques, plagioclase		31
7a–b	sandstone	quartz	hematite, chalcedony	112
7c–d	siliceous oolite	quartz, chalcedony		112
7e–f	shale	clay, quartz	pyrite, organics	112
8a–b	limestone	calcite		113
8c–d	graywacke	calcite, quartz		113
8e–f	limestone	calcite, dolomite		113
9a–b	concretion	quartz, chalcedony		114
9c–d	arkose	K-Feldspar, quartz	muscovite, hematite	114
9e–f	sandstone	glauconite, chert		114
10a–b	sandstone	quartz, calcite, collophane		115
10c–d	sandstone	quartz, calcite		115
10e–f	geode	siderite	hematite	115
11a–b	anhydrite	anhydrite		116
11c–d	gyprock	gypsum		116
11e–f	barite	barite		116
12a–b	volcanic tuff	plagioclase, opaques, glass		117
12c–d	andesite	plagioclase, hornblende, glass	clinopyroxenes, magnetite	117
12e–f	granite	quartz, biotite		117
13a–b	granite	quartz, biotite	plagioclase, microcline	118
13c–d	syenite	plagioclase, nepheline		118
13e–f	gabbro	plagioclase, augite		118
14a–b	quartz monzonite	plagioclase	sericite	119
14c–d	syenite	orthoclase, hornblende, biotite	opaques, sericite	119
14e–f	orthoclase	orthoclase	sericite, calcite	119
15a–b	anorthoclase	anorthoclase, olivine		120
15c–d	syenite	nepheline	sericite	120
15e–f	leucite	leucite	opaques, plagioclase	120
16a–d	migmatite	perthite (plagioclase, K-feldspar), quartz		121
16e–f	orthogneiss	myrmekite, quartz, plagioclase		121
17a–b	granodiorite	plagioclase, quartz, biotite	chlorite, apatite	122
17c–d	orthogneiss	plagioclase, quartz, biotite, hornblende	titanite	122
17e–f	granodiorite	plagioclase, quartz, microcline, hornblende	titanite	122
18a–b	gneiss	quartz, biotite	tourmaline, zircon	123
18c–d	granulite	biotite	zircon, opaques	123
18e–f	granite	quartz, biotite, muscovite	rutile, apatite	123
19a–b	dunite	olivine	chlorite, opaques	124
19c–d	gabbro	olivine, plagioclase	iddingsite, opaques	124
19e–f	basalt	olivine, augite	plagioclase, opaques	124
20a–d	basalt	augite	plagioclase, opaques	125
20e–f	marble	diopside, calcite		125
20g–h	blueschist	jadeite, glaucophane, riebeckite	opaques	125

Plate	Lithology or Major Mineral	Major Minerals Seen in Photographs	Minor Minerals	Page
21a–b	anorthosite	pigeonite	sericite	126
21c–d	granulite	hypersthene, quartz, biotite, plagioclase		126
21e–f	skarn	wollastonite, garnet, diopside		126
22a–b	amphibolite	hornblende, clinopyroxene, plagioclase		127
22c–d	syenite	hornblende, plagioclase, biotite	sericite	127
22e–f	blueschist	glaucophane, riebeckite	titanite, hornblende, chlorite	127
23a–b	schist	actinolite	chlorite	128
23c–d	marble	tremolite		128
23e–f	marble	tremolite	quartz	128
24a–b	schist	cummingtonite	garnet, biotite	129
24c–d	grunerite, magnetite	grunerite, magnetite	chlorite, hematite	129
24e–f	schist	talc	tremolite	129
25a–b	serpentinite	serpentine (antigorite)	magnetite, magnesite	130
25c–d	granite	muscovite, plagioclase, microcline, quartz		130
25e–f	diorite	biotite, plagioclase		130
26a–b	marble	phlogopite, calcite, diopside		131
26c–d	tonalite	biotite, chlorite, plagioclase		131
26e–f	granite	chlorite, biotite, muscovite, quartz, plagioclase		131
27a–b	epidosite	epidote, quartz		132
27c–d	basalt	plagioclase, hematite, epidote, quartz		132
27e–f	blueschist	lawsonite, glaucophane-riebeckite	opaques	132
28a–b	basalt	stilbite, opaques		133
28c–d	schist	garnet, stilpnomelane		133
28e–f	eclogite	garnet, omphacite (altered to actinolitic hornblende)		133
29a–b	schist	andalusite	sericite	134
29c–d	schist	andalusite, quartz, muscovite	sillimanite	134
29e–f	schist	sillimanite, biotite, quartz		134
30a–b	quartzite	kyanite, quartz, plagioclase	sericite	135
30c–d	quartzite	staurolite, quartz		135
30e–f	schist	chloritoid, muscovite, quartz, garnet	graphite	135
31a–b	gneiss	spinel, magnetite, biotite, quartz, plagioclase		136
31c–d	gneiss	spinel, cordierite, biotite	sillimanite	136
31e–f	gneiss	sillimanite, cordierite, biotite, quartz, plagioclase, garnet		136
32a–b	granulite	cordierite, biotite, hypersthene	zircon	137
32c–d	gneiss	corundum, sapphirine, hypersthene	biotite	137
33a–b	metabasite	scapolite, clinopyroxenes	apatite, titanite	138
33c–d	skarn	vesuvianite	calcite	138
33e–f	marble	pyrite, calcite	tremolite	138
34a–b	amphibolite	hornblende, chalcopyrite, plagioclase, quartz	biotite	139
34c–d	amphibolite	magnetite, ilmenite, biotite, quartz		139
34e–f	iron formation	hematite, quartz		139

Mineral Index

Minerals in bold are discussed in detail in this book. Key page numbers are in bold.